BUILDING OWNING & FLYING A COMPOSITE HOMEBUILT

Also by the Author from TAB BOOKS Inc.

No. 2372   *The Illustrated Buyers Guide to Used Airplanes*

BUILDING OWNING & FLYING A COMPOSITE HOMEBUILT

BILL CLARKE

TAB BOOKS Inc.
Blue Ridge Summit, PA 17214

Notice

Kevlar is a registered trademark of DuPont.

For the true craftsmen, who labor with love.

FIRST EDITION
FIRST PRINTING

Copyright © 1985 by TAB BOOKS Inc.
Printed in the United States of America

Reproduction or publication of the content in any manner, without express permission of the publisher, is prohibited. No liability is assumed with respect to the use of the information herein.

Library of Congress Cataloging in Publication Data

Clarke, Bill (Charles W.)
Building, owning, and flying a composite homebuilt.
Includes index.
1. Airplanes, Home-built. 2. Composite construction.
I. Title.
TL671.2.C55   1985   629.133′343   85-22243
ISBN 0-8306-2402-3 (pbk.)

Front cover photograph courtesy of Rex Taylor of Viking Aircraft.

Contents

	Acknowledgments	vii
	Introduction	viii
1	**What Are Composites?**	1
2	**Materials Used in Composites**	4
	Woven and Fiber Materials—Liquid Materials—Foams	
3	**Procedures and Techniques**	13
	Foam-Over-Structure—Moldless Foam Structures—Premolded Structures—Overview	
4	**Certified Airplane Engines**	21
	Availability—Used Engines—Time vs. Value—Engine Codes—Engine Applications/Specifications—A Few Other Words—Storage of Engines	
5	**Non-Certified Airplane Engines**	31
	AMI Engines—Limbach Flugmotoren Engines—VW Conversions—Factory H.A.P.I. Engines—Monnett Engines—Building a Conversion—The Combination Alternative—Rotary Engines	
6	**Propellers**	49

7 Avionics   52
Definitions—Needs—New Equipment—Reconditioned Equipment—Used Equipment—Radio Kits—Antennas—Instrumentation—Summary

8 Popular Composite Airplanes   73
Aero Mirage—Cozy—Dragonfly—Glasair—Lancer—Polliwagen—Quickie—Rand—Rutan—Sea Hawk—Silhouette—Solo—Star-Lite—W.A.R. Replicas

9 FAA Relations   100
Certification and Operation of Amateur-Built Aircraft—Addresses—Publications—Certification of Repairmen (Experimental Aircraft Builders)

10 EAA Assistance   121
Areas of Interest—Information Services—Gatherings—Local Chapters—Divisions—Assistance—History of the EAA—Membership Information

11 Safety and Hints   131
Practice Kits—Quality Control—Hints—Health—Purchasing—Flying Safety

Appendix A   GADOs and FSDOs   137

Appendix B   Manufacturers, Suppliers, and Organizations   144

Appendix C   Construction Measurements   146

Index   165

Acknowledgments

Thanks to all the manufacturers and suppliers that provided information for this book. Without their help the task would have been impossible.

A special thanks to:
Aircraft Spruce and Specialty Company for the use of their composite materials information.
Rex Taylor of H.A.P.I. Engines for information and background material about VW conversions.
The EAA for historical data and photos.
The men and women of the FAA for their guidance in selecting FARs for inclusion in this book.

Introduction

Sometime or other everyone who has ever flown entertains thoughts of owning his own airplane. But perhaps *your* thoughts don't rest with the common everyday airplane found on today's market. After all, they are all pretty much alike; yet we, as pilots, are each unique.

In looking for an airplane that suits your individual tastes and needs, consider the various composite homebuilts. These strong, durable airplanes are available in numerous configurations, with specifications to suit most tastes.

In this book you will learn what composite construction is and how it is used in modern airplanes. You will learn the terminology, materials, and methods of compositing. Then go on to read about many of the current composite airplanes, including specifications, pictures, and items of interest about each. There is a complete section about engines used for powering these homebuilt craft, including alternatives to the expensive certificated powerplants. Avionics are approached from a no-nonsense angle, and cost-saving advice is given about their needs and purchase.

Learn how you, as the builder of a homebuilt airplane, can perform preventive maintenance, maintenance, repairs, and modifications on your homebuilt. You may even perform your own annual inspection.

Study the FARs (Federal Aviation Regulations) and learn about the FAA's role in homebuilding and how they will assist you, then follow through the necessary paperwork for registering a homebuilt. Read how the EAA (Experimental Aircraft Association) helps the homebuilder, and learn about the EAA sponsored fly-ins and contests.

You will learn about the dangers in handling the various chemicals utilized in composite construction, and discover the methods used to avoid these hazards.

All the "where-to-goes" for help, plans, kits, parts, and advice are included. Also listed are the addresses of FAA offices.

In summary, this book was written to educate the potential composite homebuilder, assist in economical decision-making, and allow the side-stepping of many pitfalls found in aircraft construction and ownership.

Chapter 1

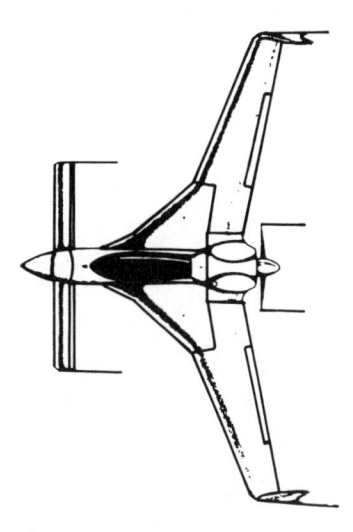

What Are Composites?

The term *composite*, when used in the context of homebuilt airplanes, refers to structures made up of more than one substance, or layers, yielding a structure stronger than any of the individual substances alone. For this discussion, *composites* will be the use of fiberglass, Kevlar, epoxies, polyesters, and various types of plastic foams.

Early examples of composite airplanes include the German gliders. These aircraft are known for their light weight, strong airframes, and smooth, sleek surfaces, all exemplifying the reasons many homebuilders desire to use composite construction.

When comparing the alternatives of construction—tube-and-fabric and all-metal—one can readily see why composite construction has become so attractive.

Tube-and-fabric construction is familiar to many of us. After all, most older airplanes are built by this means. Examples would be the Piper J3, Piper Tri-Pacers, etc. The basic framework is made up of welded steel tubing, assorted latticeworks, stringers, and shapers (most made of wood). This framework is then covered with one of several types of fabric. It is this fabric covering that controls the flow of air over the flying surfaces and allows the aircraft to fly. This fabric covering is all that is between the cabin and the elements of weather. These airplanes represent the slower, low-performance spectrum of the homebuilt field (Fig. 1-1).

All-metal construction is considered by many to be the current state of art. It is used in everything from small two-seat airplanes such as the Cessna 152, through the Beechcraft A36, and on up to the lastest airliners (Fig. 1-2).

Although nice as a finished product, and certainly faster and sleeker than the tube-and-fabric airplane, the all-metal airplane may be a more difficult project for the homebuilder. It requires special equipment, experience, and a lot of time. Although there are some fine examples of all-metal homebuilt airplanes to be seen, it is probably the hardest way to go in homebuilding.

The composite airplane, as shown by the German gliders, is sleek, therefore efficient, yet is not difficult for the homebuilder to construct. It re-

Fig. 1-1. This is a Piper PA-12, built in the late 1940s. It is a typical "tube-and-fabric" airplane. Although excellent sportplanes, such lightplanes lack the performance figures of the composites.

Fig. 1-2. This late-model Piper product is typical of current all-metal production airplanes. They are often referred to as "spamcans" by other than their owners. Although good performers, they can hardly be called sport planes; they are designed as transportation vehicles.

Fig. 1-3. One of the latest examples of the art of airplane building is this Sea Hawk composite, an amphibian. Due to the composite construction, there is no danger of corrosion, even in salt water. (courtesy Leg-Aire)

quires no special tools, equipment, or experience, yet yields easily formed contours that are impossible with tube-and-fabric design, and difficult at best with metal construction. Most builders of composite airplanes are first-time builders but *if they follow directions accurately*, their labors result in a sound airplane (Fig. 1-3).

One note, however: A composite structure is not adaptable to amateur *design* practice. If you want to build a composite, select an airplane of proven design offered by the experts, then follow their instructions to the letter.

Chapter 2

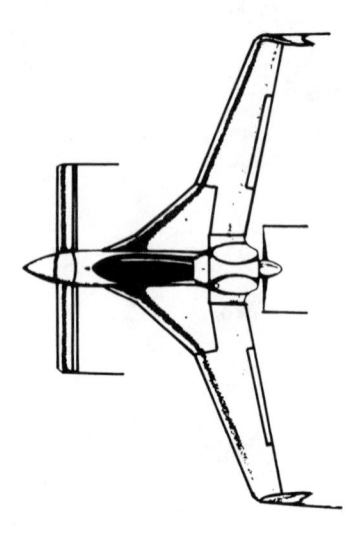

Materials Used in Composites

The materials featured in this section are those that have been used to construct proven composite designs. All builders are urged to exert *extreme* caution in the selection of proper materials for a composite project. A number of needless fatal accidents have occurred because a composite aircraft was built of non-structural materials. Anyone considering building a composite homebuilt airplane should purchase Burt Rutan's book *Moldless Composite Sandwich Homebuilt Aircraft Construction*. Burt is "the" present expert on moldless composite construction.

WOVEN AND FIBER MATERIALS

There are several types of fabrics, made of various fibers and available in different weaves. Here each is explained.

Rutan Fiberglass Cloths

The most basic structural material in building a composite aircraft is glass cloth. The use of glass in aircraft structures—particularly structural sandwich composites—is a recent development. Glass cloth is available commercially in hundreds of different weights, weaves, strengths, and working properties. Very few of these, however, are compatible with aircraft requirements for high strength and light weight. Even fewer types are suitable for the hand-layup techniques developed by Burt Rutan for the homebuilder. The glass cloth featured here has been specifically selected for the optimum combination of workability, strength, and weight. Two types—a bidirectional cloth (BID) and a unidirectional cloth (UND)—are used in aircraft structure construction. BID cloth has half of the fibers woven parallel to the selvage edge of the cloth and the other half at right angles to the selvage, giving the cloth the same strength in both directions. UND cloth has 95 percent of the glass fibers woven parallel to the selvage, giving excellent strength in that direction and very little at right angles to it. BID is generally used for pieces that are cut at a 45 degree angle to the selvage—a bias cut—which enables the builder to lay BID into contours with very little effort and provides the needed shear and torsion stiffness for flying surfaces. UND is used in areas where the primary loads are in one direction, such as wing skins and spar caps. Multiple layers of glass cloth are laminated together to form the aircraft structure. Each layer of cloth is called a *ply*.

Advanced Composite Fabrics

Advanced composite fabrics are those materials that have been used for a number of years in aerospace applications, replacing standard fiberglass fabrics. Today's materials—Kevlar, graphite, S Glass and ceramics—are now making the transition from aerospace to homebuilt aircraft.

Kevlar is an organic fiber that is yellow in color and soft to touch. It is extremely strong and tough, and about the lightest structural fabric on the market today. Kevlar is highly resistant to impact, but is rather difficult to work with for hand-layup applications. Its compressive strength is considered poor.

Graphite fibers are created by extreme stretching and heating of rayon fibers to change their molecular structure. Graphite has very low density (weight/unit volume), is very stiff (high modulus), and very strong (high tensile).

S Glass uses a different chemical formulation from standard E Glass fabrics, and is stronger, tougher, and stiffer than E Glass. One ply of S Glass can replace several plies of E Glass, which will result in a stronger and considerably lighter aircraft component.

Ceramic fabrics are the latest innovation in advanced composites. These fabrics produce laminates approaching the qualities of S Glass, plus they can withstand temperatures of almost 3000 degrees F. Ceramic cloth can produce a very lightweight and effective firewall laminate, although at this time the cost is high.

These advanced composite materials are currently being used in the production of such items as aerospace components, high-performance boats and race cars, and many revolutionary homebuilt aircraft such as the Long-EZ, Solitaire, Sea Hawk, and Q200. The performance of future homebuilt aircraft will most certainly be improved with the availability of these innovative new composite materials.

Often, the choice of the materials to use for a laminate is difficult because of the required properties. One must consider the advantages of one material over another, and the anticipated performance. S Glass is about 30 percent stronger and 15 percent stiffer than E Glass. It has 20-25 percent of the stiffness of graphite and is as strong, but it is heavier. S Glass, though, has only half the strength and stiffness of Kevlar and twice the weight.

Kevlar, on the other hand, is 40 percent stronger and 25 percent lighter than graphite but has only half the stiffness of graphite. Sometimes, blending different advanced composite fabrics in a laminate can achieve the proper balance of stiffness, strength, and weight. Use the following chart to help you decide which of these new fabrics is best in your particular application:

	Best				Worst
Cost	E Glass	S Glass	Kevlar	Graphite	Ceramic
Weight	Kevlar	Graphite	S Glass	E Glass	Ceramic
Stiffness	Graphite	Kevlar	S Glass	Ceramic	E Glass
Heat resistance	Ceramic	S Glass	E Glass	Kevlar	Graphite
Toughness	Kevlar	S Glass	E Glass	Ceramic	Graphite
Impact resistance	Kevlar	S Glass	E Glass	Ceramic	Graphite

Bidirectional Woven Kevlar

Kevlar 49 Aramid fiber was introduced commercially in 1972 and is the DuPont registered trademark for this new high-strength, high-modulus organic fabric. It combines high tensile strength (43,000 psi) and high modulus (19 million psi) with light weight and toughness superior to other reinforcing fibers for plastics. It is available in yarns and rovings that meet all FAA requirements for flammability. It shows no degradation in jet fuel, lubricating oils, water, salt water, or high humidity. At cryogenic temperatures (−320 degrees F.) performance is excellent with essentially no embrittlement or degradation of fiber properties.

Kevlar 49 can offer both a significant weight saving and improved stiffness versus glass, in addition to superior vibration damping and good impact resistance. A kayak made with Kevlar 49, for example, weighs about 18 pounds while the weight of a comparable boat made with glass would be over 30 pounds. The advantages over glass in small aircraft are similar—weight savings and improved impact resistance. Kevlar 49 is used in a number of parts on the Lockheed L-1011 because of weight savings of up to 30 percent compared to similar parts made of glass.

One unusual benefit of Kevlar is its "quietness." A cowling made of Kevlar will be quieter and less sensitive to engine vibration than its glass or graphite counterpart.

Although all of the processes used in combining resins with glass fiber are adaptable to Kevlar 49 with little or no modification, it has been found that the vinyl ester type system is most compatible. The use of polyesters is not recommended because of poor bonding with Kevlar.

There are many fabric weaves available using Kevlar 49 fiber, but the one most commonly used in the fabrication of aircraft components has a weight of 5.1 oz. per square yard, with a scoured finish.

Unidirectional Kevlar

KS-400 is a unique unidirectional reinforcing material that combines the benefits of Kevlar 49 with Owens-Corning Fiberglass S-2 Glass. KS-400 is designed for use in laminates that must have high modulus, or stiffness, and tensile strength combined with low weight or density. The material is specifically constructed to enhance properties of wet-out behavior, bondability, and impact strength in a hand laid-up composite. As compared with woven fabrics of pure Kevlar 49, KS-400 should produce laminates with better fiber-resin ratios and superior stiffness.

The integrity of KS-400 is maintained through very fine, adhesive-coated Dacron fill yarns that are bonded to, but not interwoven with, the unidirectional fibers. Only vinyl esters or epoxies should be used as impregnating resins. The presence of the S-2 Glass enables the fabricator to determine visually when the material is properly wet-out. Resin bond is far superior to the glass than to the Kevlar. This means increased resistance to delamination. Also, the presence of the S-2 Glass, even in such a relatively small amount, increases the impact strength of the laminate significantly.

Unidirectional Woven Graphites

Woven graphite is a fabric introduced in recent years that has become an excellent alternative to fiberglass and Kevlar— only mils thick with great strength. In addition to its great strength, graphite fabric also has very low density and is very stiff. Although it is quite costly, the material saving is appreciable since only one course (layer) of graphite is required for three or four of fiberglass. It cuts considerably easier than Kevlar. Graphite "pre-pregs," which are standard graphite weaves impregnated with either polyester or epoxy resins, have been used by major manufacturers to cut production time on composite parts; however, the required equipment and precise production controls for proper cure of pre-pregs make them difficult to adapt to homebuilt applications. The excellent qualities of the graphite fabric itself give it an immediate waiting market in the aircraft building field.

Unidirectional Graphite

Construction is such that the fibers are oriented in a straight or linear manner with no twist and can be maintained in that condition while being impregnated by hand. The fabric is formed from rovings or "tows" of fibers similar to those used in making woven fabric. These fibers are locked into position by very fine fill (cross machine direction) fibers which are encapsulated with an adhesive that is compatible with common impregnating resins. These fill fibers—and the encapsulating adhesive—will be visible in any clear resin. The resulting "pattern" is normal and should not be interpreted as poor wet-out of the reinforcing fibers.

For hand-layups, resin can be applied by spray, brush, or pouring action. A short-napped paint roller is recommended for spreading the resin. Excess resin should always be rolled out in a direction parallel to the graphite fibers. The final or external layer should be applied with the fill yarns facing down (unexposed). This fabric is satisfactory for use with epoxies, polyesters, vinyl esters, and other resins. The fabric is rolled with a polyethylene interliner to maintain cleanliness.

Carbon or Graphite Fibers

High-strength (470,000 psi) carbon fibers are used as reinforcement in high-performance structural composites for aircraft applications and recreational and industrial products. Carbon fiber filaments are finer than a human hair. These filaments are bundled together to make a fiber of 3,000, 6,000, or 12,0000 filaments, which is called a *tow*. The tow is sized with an epoxy-compatible material to improve the handling characteristics. It is then wound on a cardboard core holding four to six pounds of fiber. It appears that the 6,000 tow will prove most practical for homebuilding applications. The 3,000 tow is used primarily by weavers. The 12,000 tow is difficult to wet out, but it can be done by diligent brushing.

S Glass

The chemical formulation of S Glass differs from that of standard E Glass. S Glass is 30 percent stronger and 15 percent stiffer than E Glass, and retains these properties up to 1500 degrees F. S Glass is also considerably tougher than an equivalent amount of E Glass.

Ceramics

A number of fiber manufacturers have recently developed the technology to produce continuous filaments of inorganic (mineral) fibers which can be woven into cloth. These fabrics produce laminates approaching the mechanical properties of S Glass and they can withstand temperatures of almost 3000 degrees F. Their handling is similar to fiberglass. Ceramic fabrics are currently in use as lightweight fire protection barriers, replacing heavier metal on many of today's newest aircraft. At the present time, ceramic cloth is offered in only one style. The only drawback to this excellent new material is its extremely high cost. For further information I recommend the booklet *Advanced Composite Fabrics*, published by Hexcel Corp. Complimentary copies are available from Aircraft Spruce & Specialty Company.

Standard Fiberglass Cloth

Fiberglass cloth is exactly what the name says—glass. Fine fibers are spun from molten glass marbles, gathered into yarn, and woven into a strong yet supple glass fabric. It can be folded, rolled, or draped, like any other loosely woven fabric— but can be chemically transformed into solid sheets of tremendous strength.

Standard fiberglass cloth is seen in two weaves. The weave pattern describes the manner in which the warp yarns and the filling yarns are interlaced in the fabric.

Plain weave consists of one warp end woven over and under one filling pick. Plain weave is generally characterized by fabric stability with minimum pliability except at low fabric counts (Fig. 2-1).

Crowfoot weave is constructed with one warp end weaving over three and under one filling pick. It is characterized as being more pliable than either plain or basket weaves, having conformability to complex or compound-curved surfaces, and making possible the weaving of higher counts than plain or basket weaves (Fig. 2-1).

Unidirectional Fiberglass

Unidirectional fabric is constructed with Owens-Corning Fiberglass S-2 Glass, offering outstanding strength-to-weight ratio, superb glass-resin ratio control to minimize probability of resin-

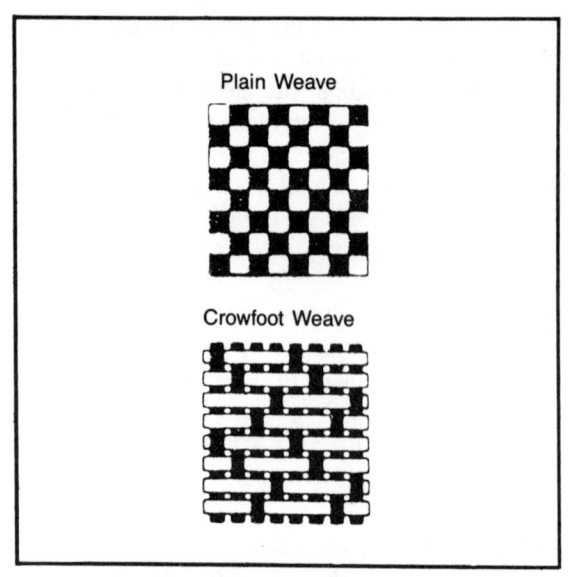

Fig. 2-1. Examples of plain weave and crowfoot weave. (courtesy Aircraft Spruce and Specialty Company)

rich and/or resin-dry areas, handlability without distortion, and exceptionally high impact resistance. The integrity of S-500 is maintained through very fine adhesive-coated fill yarns that are bonded to, but not interwoven with, the unidirectional fibers of the S-2 Glass. The fill yarns are spaced approximately 1.5 inches apart. The use of short-nap paint rollers is suggested, rolling under pressure, always parallel to the fibers. Use with epoxies, vinyl esters, and polyesters.

Glass Mat

Glass mat is 100 percent fiberglass in a nonwoven state. It is used for bulk "build-up" in molding and fabricating components. It's also useful for filling holes and badly damaged parts.

Pre-Cured Fiberglass Laminate

Fully cured glass cloth laminates with polyester resin binder systems are used in commercial jet aircraft for ceilings and as cargo liner material. They are used as skin material in the construction of some sailplanes. These laminates are compatible with composite construction methods and bond well when epoxied to wood, foam, paper, and most low modulus materials. They do not bond to aluminum and other metals. They should be excellent for use as spar caps.

LIQUID MATERIALS

Polyesters, epoxies, curing agents, etc., are all in liquid form, and are applied to the fabrics in what is called *wet-out*.

Polyester Resins

Polyester resins are *hygroscopic* (they draw moisture from the air). There are two types of resin, and one or both types may be required, depending on the application.

Type "A" resin, *surface resin*, has a small amount of wax in it, which comes to the surface and forms a barrier against air-bound moisture. This permits the resin to cure completely—the surface is hard and easily sanded.

Type "B" resin *bond coat*, does not have any wax content. As a result, the surface stays a little tacky, as the surface cure is being inhibited by moisture. This tacky surface provides excellent adhesion between coats. Bond coat resin is therefore recommended for the first coat of resin to fill the weave of the glass cloth, for bonding fiberglass to plywood or other surfaces, and for multi-layers of glass cloth.

Iso Resin

Iso resin is highly flexible and has excellent adhesion to metal, wood, concrete, fiberglass, and other "hard-to-adhere" surfaces. It is compatible with most fuels. Iso is a wax-free resin and must be overcoated with Type "A" surfacing resin to obtain a surface cure.

Surface Curing Agent

A surface curing agent is used in polyester resin to improve sanding properties. Added to the bond coat, it will provide the same sanding properties as in Type "A" surface resin. The surface curing agent is used in addition to a catalyst.

Epoxy Systems

Just as fiberglass products are extremely varied, epoxy systems also differ greatly in their working properties—some are thick, slow-pouring liquids while others have a consistency similar to water. Some allow hours of working time and others harden almost as fast as they are mixed. The RAF (Rutan) epoxy system was formulated to meet the rigid specifications set forth by Burt Rutan at the outset of the VariEze program in 1975. Many builders still prefer the working properties of the RAF system, although most new builders are using the low-toxicity Safe-T-Poxy products introduced in 1980. Both these systems provide the combination of easy workability and the strength required for composite aircraft construction.

Safe-T-Poxy

As hundreds of VariEze projects progressed, one persistant problem plagued many builders—sensitivity to the chemicals in the epoxy system (although the RAF system was by far the least toxic of all the systems available). A number of builders developed severe cases of dermatitis and were forced to abandon their projects. Applied Plastics, manufacturer of the RAF system, was keenly aware of the problem and concerned for the safety of the builders. A full-scale research project was launched beginning with data accumulated by their research director years before. In only eight months, they were able to develop a system that is so low in toxicity that incidental contact is not hazardous—a tremendous breakthrough.

In addition to the obvious advantages of being safe to use, this new Safe-T-Poxy system, which uses only one speed of hardener, has other desirable characteristics. It has unusually low water absorption and will cure to a dry surface at high humidity, at room temperature, with very low exotherm. Since the material actually repels water, it cannot have a surface stickiness as caused by humidity in the early system. Other redeeming features builders will find in using it are:

- No fillers to settle out.
- Superior wetting of the fiber/cloth with little drain-out of resin.
- Lower density resin that yields a lighter laminate.
- A 35 percent reduction in exotherm with no change in cure rate.
- Cost competitive with the older established systems.

With further respect to the safety features, previous resins contained glycidyl ethers or various solvents to reduce viscosity for better handling. Most of these are now NIOSH restricted. The new resin contains none of these. Additionally, most epoxy curing agents contain amines that will severely irritate the skin. When these are measured by a Draise Skin Irritation Test, in almost all epoxy cases they will yield a rating of 5—or worse. The new hardener rates a 0 in this test.

Specifications:

Mixing error:	+/−5 percent.
Pot life:	45 minutes at 77 degrees F.
Flash point:	320 degrees F. (resin).
	172 degrees F. (hardener).
Tack-free:	4 hours at 77 degrees F.
Cure:	10 hours at room temperature.
Storage life:	One year in closed container.

Poxipol II

The epoxy system used for the Polliwagen is basically the same formulation as Safe-T-Poxy with slightly higher viscosity.

RAF Epoxy System

Some builders who are not sensitive to epoxy toxicity still prefer working with the RAF system as originally developed for use on the VariEze. This system offers two speeds of hardeners: RAFS (slow curing) and RAFF (fast curing). Both the RAFS and RAFF hardeners use the same RAF resin.

Specifications:

Pot life:	20-45 minutes (RAFF).
	1-2 hours (RAFS).
Cure:	3-6 hours (RAFF).
	10-16 hours (RAFS).

FOAMS

There are currently five different types of rigid, closed-cell foams being used in proven designs.

Styrofoam FB

Styrofoam FB is a low-density (2 lb./cu.ft.), large-cell, fire-retardant material. It varies in color from white to blue. The large-cell type provides better protection from delamination than the more commonly used small-cell blue styrofoam. It cuts smoothly with a hot wire for airfoil shapes. Do not confuse styrofoam with expanded polystyrene, which is the type seen in the average picnic cooler. The compressive strength of polystyrene is too low for use in aircraft structures. Also, it dissolves in most solvents and fuels.

Orange/Blue Styrofoam

Orange/blue styrofoam, getting its name from its color, is of low density (2 lb./cu.ft.), and has been used for years as flotation for boat docks and for other marine uses. Commonly called a *buoyancy billit*, this foam's applications and costs are similar to the FB (blue/white) type. Its prime advantage is its availability in the large billit sizes. This foam is now used in the Q2 design, and should be used with Safe-T-Poxy only. RAF type epoxy tends to bleed into the foam, which will result in a dry layup.

Clarkfoam

Clarkfoam is a medium density (4 lb./cu.ft.) material, white in color, that is available in sheets varying from 1/8 inch to 4 inches in thickness. Its primary use is as a center in laminated sections.

Klegecell

Klegecell is a standard foam meeting FAA regulations for fireproofing. It is available in many densities from 2 lb. to 15 lb./cu.ft. It is heat-moldable at a temperature of 200 degrees F.

Urethane

Urethane foam is low density (2 lb./cu.ft.), small-cell, colored green or tan. It's used extensively in the construction of fuselages and fuel tanks. It is completely fuelproof, and is easy to carve and contour with a large knife.

Urethane Polyester

Urethane polyester foam is of medium and high density (4 to 18 lb./cu.ft.), white, and small-celled. It can be readily cut and carved, and has limited heat forming characteristics. Its fuel compatibility makes it suitable for sandwich-type fuel tank construction. Its uniform surface, excellent compressive strength, and low cost make it an ideal replacement for some types of PVC foam applications.

PVC

PVC foam is medium to high density (3 to 15 lb./cu.ft.) and is used in fuselage bulkheads and other areas where high compressive strength is required. The tan-colored PVC foam has excellent workability and can be used in place of comparable blue PVC and white Clark foams specified in the Long-EZ plans.

Liquid "X-40" Foam (two-component Polyurethane Foam)

This system consists of two components—"X-40" resin and "X-40" catalyst. When the resin and catalyst are mixed in equal volumes, they expand into a rigid closed-cell foam of 2 lb. density. Thorough mixing of the two components is essential. "X-40" foam expands approximately 40 times its liquid volume. Cured foams can be easily trimmed, cut, and shaped with common woodworking tools. "X-40" foam contains a highly reactive agent and is classified as a toxic material. It is combustible, a strong skin sensitizer, and an eye irritant. Avoid contact with the skin.

Poly-Cel One

Poly-Cel One is a one-component polyurethane foam that requires no mixing. It dispenses like shaving foam from an aerosol can, then sets up to a rigid closed-cell foam. It adheres permanently to almost any surface. It does not shrink, dry out, or become brittle with age.

Fillers, Tapes, and Primers

In the completion of any project there will always be rough edges, mistakes, etc., that must be covered or filled. In addition, there will be joints to be made, sealed, and covered. The following is a description of the various products available for these jobs.

Glass Bubbles

These bubbles are actually hollow glass spheres. Because the high-quality glass is very crush-resistant, the glass bubble/resin mix is much stronger, stiffer, and water-resistant than any foam made by chemical means. These bubbles displace four to six times their weight in most resins, and improve the handling characteristics of the base resin. They have a low bulk density and are nontoxic.

Mix resin and hardener as directed, then fold in the glass bubbles. Upon cure, a strong, low-density product results that is easy to sand and file. It may be shaped to form compound angles and curves.

The term "micro" was applied to the mixture of microspheres and epoxy early in the development of composite structures. Although microspheres have been replaced by glass bubbles, the word "micro" is still commonly used to reference the mixture. "Micro" is used to fill voids and low areas, to glue foam blocks together, and as a bond between foams and glass cloth. Micro is used in three consistencies:

- ☐ "Slurry," which is a one-to-one (by volume) mix of epoxy and glass bubbles.
- ☐ "Wet micro," which is about two to four parts glass bubbles (by volume) to one part epoxy.
- ☐ "Dry micro," which is a mix of epoxy with enough glass bubbles to obtain a paste that will not sag or run (about five parts to one by volume).

In all instances, bubbles are added to completely mixed epoxy resin and hardener. Wet micro is used to join foam blocks and is much thicker than slurry (it has the consistency of honey), but can be brushed. Dry micro is used to fill low spots and voids and is mixed so that it is a dry paste and will not sag. Apply with a putty knife.

Flocked Cotton Fiber

Flocked cotton fiber is structural resin filler. The mixture of cotton fiber and epoxy is referred to as *flox*. The mixture is used in structural joints and in areas where a very hard, durable buildup is required.

Flox is mixed in much the same way as dry micro, but only about two parts flock to one part epoxy is required. Mix in just enough flock to make the mixture stand up. If "wet flox" is called for, mix it so it will sag or run.

Flox is often used to reinforce a sharp corner. Paint a light coat of pure epoxy inside the corner, then trowel flox in to make a triangular support. The flox corner is done just before one glass surface is applied for a wet bond to one surface.

Bondo

Bondo is an automobile body filler used extensively in composite construction to hold jig blocks in place and for other temporary fastening jobs. It hardens quickly and can be chipped off without damaging the fiberglass. The color of the mixture is used to judge how fast it will set. As more hardener is added, the brighter in color the mixture becomes and the faster it hardens.

The original #211 series automotive plastic filler has been improved using a microsphere formula producing a lightweight filler that sands easier and spreads smooth. It can be shaped days later. It is used extensively in composite construction.

Ultralite

Ultralite is a formulation of polyester resin, talcs, and microspheres. It is used as a lightweight filler on metal and fiberglass. It works easily and sands fast.

Feather Fill

Feather Fill is a sprayable polyester filler/primer used for the filling of small surface irregularities such as scratches, blemishes, and exposed fiberglass threads before final sanding and painting. It adheres to bare metal, plastic filler, and fiberglass with minimal surface preparation, and cures ready to sand and paint in 45 to 60 minutes. Any type of finish (lacquer, enamel, acrylic, etc.) may be applied over Feather Fill.

Micro-Putty

Micro-Putty is a flexible, lightweight plastic filler containing 60 percent (by volume) glass bubbles and non-bleeding chromate pigments in a polyester resin cured with a cream hardener. It will adhere to most clean surfaces (fiberglass, wood, metals, etc.). Micro-Putty sets in 10 to 15 minutes to a smooth surface.

Sterling Filler

Sterling Filler is a very durable product, formulated for aircraft use, that can be applied by spraying and has excellent adhesion qualities. It is used for fabric weave filling, and can be sanded one hour after application.

Peel Ply

Peel Ply is a layer of 2.7 oz. Dacron fabric strips or tape laminated into a layup as if it were an extra ply of glass. The peel coat wets out with epoxy like glass cloth and cures along with the rest of the layup. However, Dacron does not adhere structurally to the glass; when peeled away, it leaves a surface that is rough enough to allow additional glass-to-glass bonding (layups) without the need for additional sanding.

Fiberglass Tapes

Fiberglass tape is a fibrous glass reinforcement designed to furnish exceptionally high directional strength. It is manufactured from parallel strands of glass rovings which are held together by a fine woven cross-thread. The placement of the cross-thread is such that the parallel rovings do not wander or have a tendency to cross over each other. This 100 percent unidirectional tape can be used instead of unidirectional fabric for spar caps, wings, elevators, etc. It contours well and cuts building time considerably. It can be used with either polyester resin or epoxy systems.

#77 Spray Adhesive

Manufactured by 3M, this spray adhesive is ideal for the lamination of styrofoam sheets. The laminated blocks can then be hot-wire cut.

Chapter 3

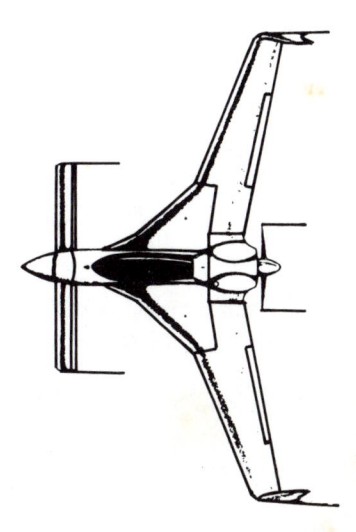

Procedures and Techniques

The beauty of composite construction is its high strength-to-weight ratio. The composite structures' high bending and buckling strengths are attained by the high section moduli of composite construction, thereby eliminating the need for many ribs, bulkheads, and other supports found in "normal" airframe structures.

There are three basic methods, or types, of composite aircraft construction. Although each method uses similar materials, each is different in the actual application of these materials, and each will utilize completely different airframe construction. Each method is explained in the following pages.

FOAM-OVER-STRUCTURE

As the name *foam-over-structure* implies, this method of composite construction is the application of foam materials over an underlying basic structure. It is the foam that gives the necessary aerodynamic lines to the airplane for flight. This can also be the basis of aesthetically pleasing aircraft lines.

Perhaps the most well-known homebuilt airplanes constructed with the foam-over-structure method are the Rand series—the KR-1 and KR-2.

The basic airframe structure of the Rand airplanes is a box-type wood frame. This frame is the actual weight-bearing structure; the foam is applied over the airframe only for shaping purposes. This system has proven popular with many first-time composite homebuilders, as the builder feels more comfortable with the rather basic wood frame structure, and knows that the foam actually gives no additional strength, just aerodynamic and decorative shaping.

The foam is glued to the wood structure, then shaped by cutting, carving, or sanding. After the final shape has been reached, the foam is covered with a synthetic material. Originally Dynel fabric was used; however, it is now difficult to find, and most foam-over construction utilizes glass cloth as a substitute. Either way, the method of application is the same: The resin is brushed or rolled onto the shaped structure, then the cloth is applied to it. After the cloth has been applied, more resin will be applied to fill the weave. As this covering serves no structural needs, only one layer of cloth is used.

As mentioned above, the shaping is easily done by cutting, carving, and sanding. Since the basic airframe is wood, and handles all the stress loads, this allows a considerable amount of freedom to the builder in the actual shaping of the aircraft. With the ease of shaping in mind, and the ever-present "WWII fighter pilot fever" that lurks in many homebuilders, the foam-over system has become a natural for the designers and builders of scale-sized fighter planes.

WAR Replicas produces 1/2-scale kits/plans for the construction of homebuilt airplanes that resemble the great WWII fighters. All WAR planes use the same basic wood airframe structure. The final contour shaping is done by cutting, carving, and sanding the foam that is applied over the basic wood airframe.

A properly finished WAR plane is wondrous to see. At a distance, or in the air, these diminutive fighters look like the real thing. When seen up close, they make the beholder wonder if he is having vision problems: "I always thought those old warplanes were bigger than this!"

MOLDLESS FOAM STRUCTURES

The building technique used in moldless construction is based on shaped foam covered with layers of glass (or other) fabric. It is these multiple layers, or *layups*, that bear the load of the structure. The foam is the means of shaping the layers of cloth. There are no internal supports within a structure, such as the wing (i.e., ribs), only a spar cap.

Moldless construction is a simple method, and requires only the most elemental tools:

- ☐ Sharp butcher knife.
- ☐ Sanding block.
- ☐ Course and fine files.
- ☐ "Hot wire" cutter.
- ☐ Dust mask.

The last item on the list, although not very glamorous, could well *save your life*. Buy and use the best dust mask you can find. This will keep the air you breathe clear of the many fine particles of composite materials that will abound in your work area.

The entire secret in moldless construction can be summed up in one statement: *Make perfect cores*. In other words, a foam core that is flawless will result in a completed layup that is also flawless. It is much easier to shape foam than it is to reshape the cured layers of glass cloth. The latter is extremely hard, and since it is weight/stress bearing must be sanded or reshaped very carefully. To reduce the thickness of the layups can dangerously reduce the strength of the overall structure.

Preparing a Foam Core

Blocks of foam are shaped by cutting. Normally this cutting is accomplished by the use of a "hot wire" cutter. The hot wire cutter is a device that utilizes a nichrome wire under tension, heated by the controlled application of electricity, to cut and shape the foam.

A good example of a hot wire cutter is one made from a hunting bow. The nichrome wire is used in place of the bowstring, thereby remaining under tension.

Using a hot wire cutter is simplicity itself. Apply electricity to the wire, causing the wire to heat up. The source of electricity can be a car battery or (preferably), a variable source. This variable source will allow for accurate temperature control of the cutting wire. Variable sources are available commercially, or can be substituted by the use of a model train transformer. Two to four amps will be required. Initial adjustment is made by allowing the wire to glow dull red, then cutting back slightly on the power until the wire no longer glows. A properly operating hot wire will cut through foam easily with only a small bit of noticeable drag, giving a slight "sizzle" sound and producing long, very fine plastic hairs from the wire at the end of a cut. This will allow about one inch of travel through the foam every four to six seconds with light pressure on the hot wire cutter.

Always practice on scrap foam pieces with the hot wire cutter before you actually attempt a cut that will count.

When cutting foam blocks it is necessary that the foam be secured to prevent any motion. Motion could cause an inaccurate cut. Remember, the core must be *perfect*.

The actual cutting will be done using templates. These templates are guides that the wire cutter rides on in order to produce a perfectly contoured shape. The templates are temporarily attached to the foam blocks that are being cut. Due to the spacing of these templates, and the size of some of the cuts that must be made, cutting can often be a two-person job.

Suppose you are not Mr. Perfection and (like most homebuilders), make a few mistakes when cutting with the hot wire. Don't fret; repair is very easy. The application of microballoon (glass bubble) micro-fill will aid in filling small voids. Later sanding will completely remove any blemishes.

Glassing the Core

Using unidirectional and bidirectional cloth, as needed for the particular part you are covering, the application will be made in layers. It is the bonding of these layers that gives the composite structure its cumulative strength.

The actual layup procedure consists of applying resin to the foam core, laying on glass cloth, squeegeeing more resin into the glass cloth, then applying the next layer of cloth. This drill is repeated until the required number of layers have been applied (Fig. 3-1).

PREMOLDED STRUCTURES

The latest rage among composite homebuilders is the advent of the premolded structure—or, in fact, the complete premolded airplane. The basis of manufacturing structures with this method is that each part will be identical with the last in size, shape, and structural integrity. Compared to moldless structures, there is a weight advantage, as the premolded structure is hollow. This could allow a savings of better than 50 pounds in the wings alone.

The premolded structures are built assembly-line fashion by the use of molds. Briefly, the fiberglass is laid up inside a mold, cured, and removed. The resulting unit can be as complete as the builder wants. The builder (factory) may include such items as dimpling (for use as drilling marks) to aid the homebuilder when he starts attaching these structures/parts together (Fig. 3-2).

Many premolded parts or complete structures are available for the "moldless" aircraft. Although somewhat more expensive at initial purchase, when compared to the raw materials, their use can be a real timer-saver—particularly when compared to the time spent in cutting, shaping, and laying up a structure. Companies and individuals producing premolded parts often advertise in *Sport Aviation* and other aviation publications directed towards the homebuilder.

Complete composite airplanes are available as kits for the homebuilder. These airplanes are growing in popularity as they require less work to build than a foam-over or moldless craft. In fact, they closely resemble an overgrown plastic model airplane kit (Fig. 3-3). The majority of the airframe work the builder will be doing is the bonding of sections together. However, don't get the impression that you're going to build an airplane in a weekend or two, just because some or all of the airplane is premolded.

The actual building of a premolded airplane requires considerable time-consuming fitting, cutting, and fastening (Fig. 3-4). All of the "plumbing" such as control systems, electrical wiring, etc., will be as difficult and complicated as for any other type of homebuilt composite-constructed aircraft.

OVERVIEW

Now that the basics of each type of composite construction have been covered, where does that leave you? I strongly recommend your attendance at one or more EAA (Experimental Aircraft Association) Chapter meetings. This will give you the opportunity to meet and talk with current and past builders. Do this, and if someone is building

REPRODUCED BY PERMISSION OF THE RUTAN AIRCRAFT FACTORY, INC.

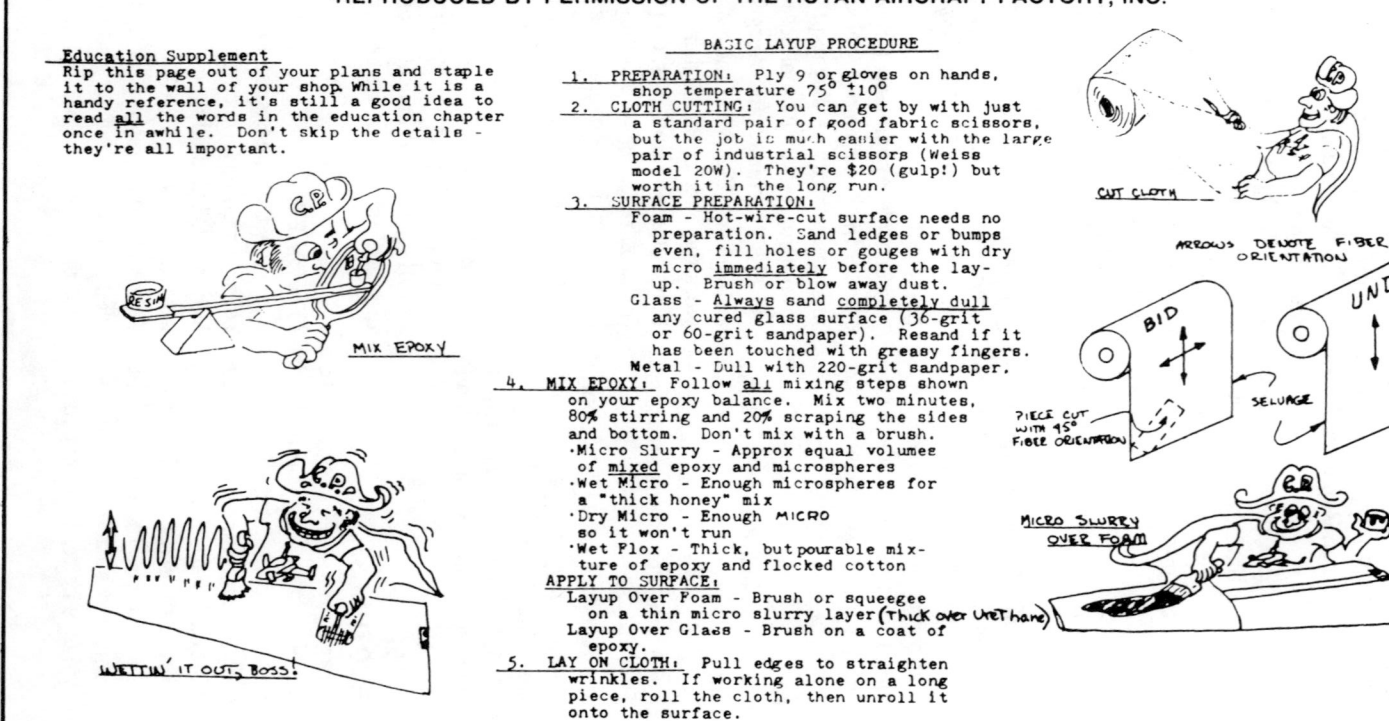

Education Supplement
Rip this page out of your plans and staple it to the wall of your shop. While it is a handy reference, it's still a good idea to read all the words in the education chapter once in awhile. Don't skip the details - they're all important.

BASIC LAYUP PROCEDURE

1. PREPARATION: Ply 9 or gloves on hands, shop temperature 75° ±10°
2. CLOTH CUTTING: You can get by with just a standard pair of good fabric scissors, but the job is much easier with the large pair of industrial scissors (Weiss model 20W). They're $20 (gulp!) but worth it in the long run.
3. SURFACE PREPARATION:
   Foam - Hot-wire-cut surface needs no preparation. Sand ledges or bumps even, fill holes or gouges with dry micro immediately before the layup. Brush or blow away dust.
   Glass - Always sand completely dull any cured glass surface (36-grit or 60-grit sandpaper). Resand if it has been touched with greasy fingers.
   Metal - Dull with 220-grit sandpaper.
4. MIX EPOXY: Follow all mixing steps shown on your epoxy balance. Mix two minutes, 80% stirring and 20% scraping the sides and bottom. Don't mix with a brush.
   •Micro Slurry - Approx equal volumes of mixed epoxy and microspheres
   •Wet Micro - Enough microspheres for a "thick honey" mix
   •Dry Micro - Enough MICRO so it won't run
   •Wet Flox - Thick, but pourable mixture of epoxy and flocked cotton
APPLY TO SURFACE:
   Layup Over Foam - Brush or squeegee on a thin micro slurry layer (thick over urethane)
   Layup Over Glass - Brush on a coat of epoxy.
5. LAY ON CLOTH: Pull edges to straighten wrinkles. If working alone on a long piece, roll the cloth, then unroll it onto the surface.

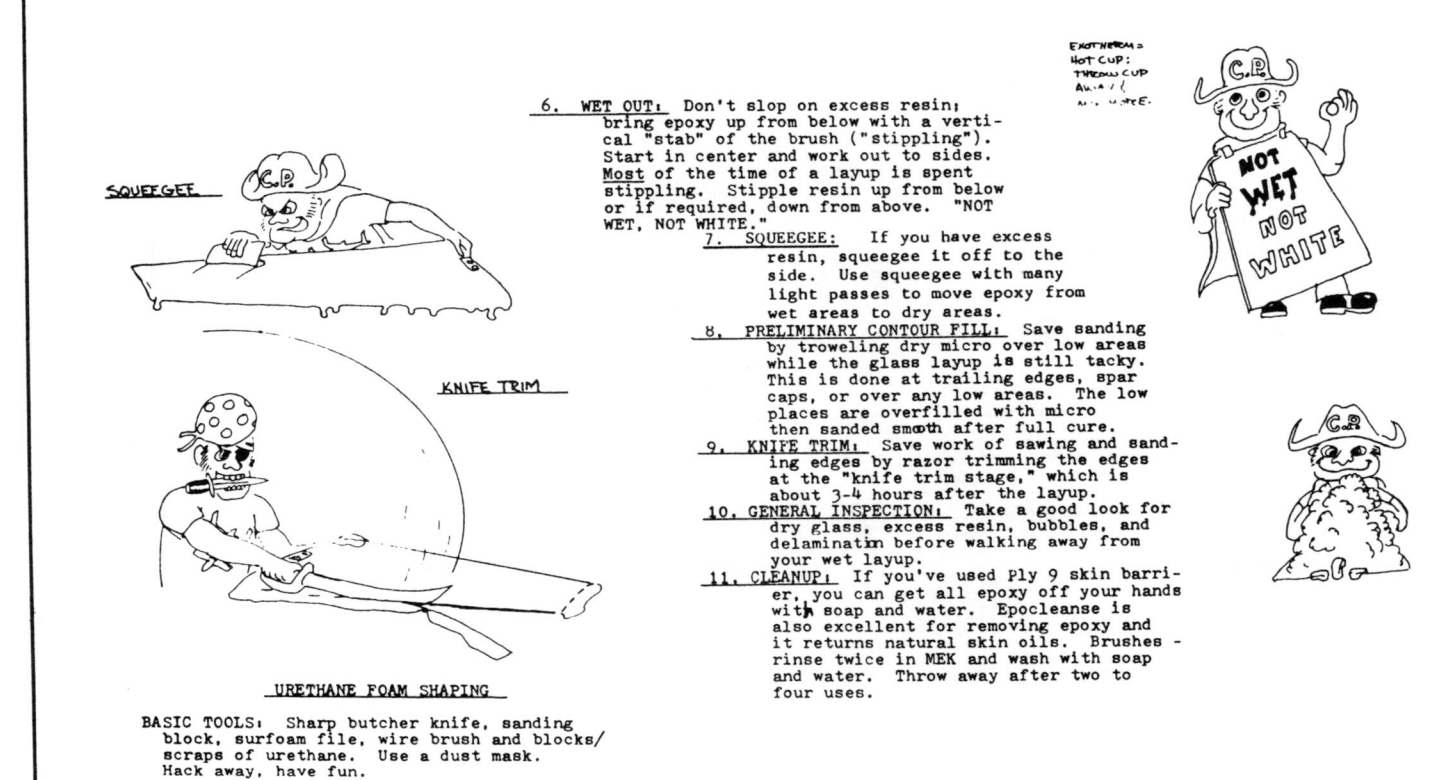

6. **WET OUT:** Don't slop on excess resin; bring epoxy up from below with a vertical "stab" of the brush ("stippling"). Start in center and work out to sides. Most of the time of a layup is spent stippling. Stipple resin up from below or if required, down from above. "NOT WET, NOT WHITE."

7. **SQUEEGEE:** If you have excess resin, squeegee it off to the side. Use squeegee with many light passes to move epoxy from wet areas to dry areas.

8. **PRELIMINARY CONTOUR FILL.** Save sanding by troweling dry micro over low areas while the glass layup is still tacky. This is done at trailing edges, spar caps, or over any low areas. The low places are overfilled with micro then sanded smooth after full cure.

9. **KNIFE TRIM.** Save work of sawing and sanding edges by razor trimming the edges at the "knife trim stage," which is about 3-4 hours after the layup.

10. **GENERAL INSPECTION.** Take a good look for dry glass, excess resin, bubbles, and delamination before walking away from your wet layup.

11. **CLEANUP.** If you've used ply 9 skin barrier, you can get all epoxy off of your hands with soap and water. Epocleanse is also excellent for removing epoxy and it returns natural skin oils. Brushes - rinse twice in MEK and wash with soap and water. Throw away after two to four uses.

BASIC TOOLS: Sharp butcher knife, sanding block, surfoam file, wire brush and blocks/scraps of urethane. Use a dust mask. Hack away, have fun.

Fig. 3-1. Here in one shot are the basics of moldless construction. (courtesy Rutan Aircraft Factory, Inc.)

HOT WIRE CUTTING STYROFOAM

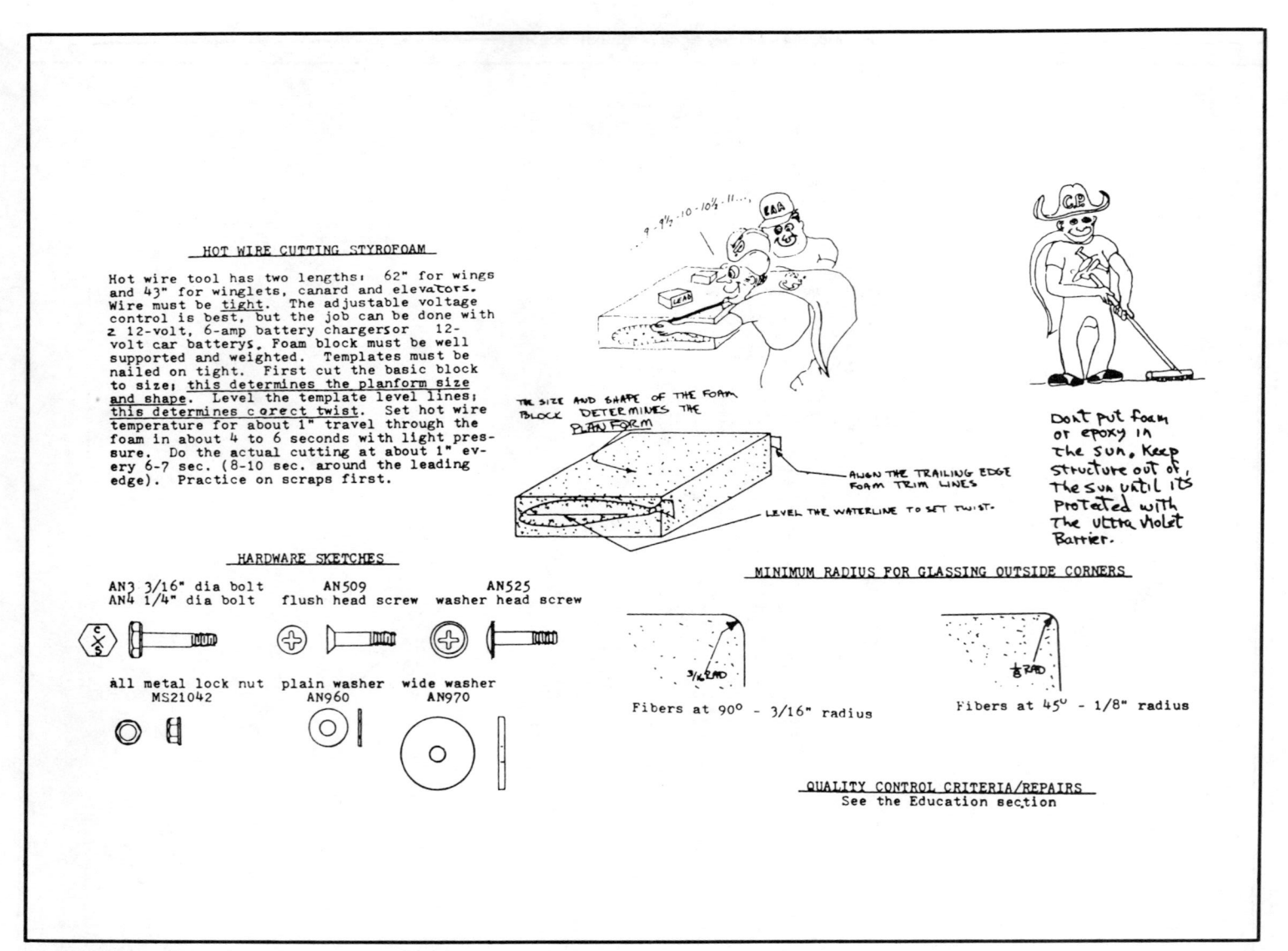

Hot wire tool has two lengths: 62" for wings and 43" for winglets, canard and elevators. Wire must be <u>tight</u>. The adjustable voltage control is best, but the job can be done with 2 12-volt, 6-amp battery chargersor 12-volt car batterys. Foam block must be well supported and weighted. Templates must be nailed on tight. First cut the basic block to size; this determines the planform size and shape. Level the template level lines; this determines corect twist. Set hot wire temperature for about 1" travel through the foam in about 4 to 6 seconds with light pressure. Do the actual cutting at about 1" every 6-7 sec. (8-10 sec. around the leading edge). Practice on scraps first.

Don't put foam or epoxy in the sun. Keep structure out of the sun until its protected with the ultra violet Barrier.

ALIGN THE TRAILING EDGE FOAM TRIM LINES

LEVEL THE WATERLINE TO SET TWIST

THE SIZE AND SHAPE OF THE FOAM BLOCK DETERMINES THE PLANFORM

HARDWARE SKETCHES

AN3 3/16" dia bolt
AN4 1/4" dia bolt

AN509 flush head screw

AN525 washer head screw

plain washer AN960

wide washer AN970

all metal lock nut MS21042

MINIMUM RADIUS FOR GLASSING OUTSIDE CORNERS

Fibers at 90° - 3/16" radius

Fibers at 45° - 1/8" radius

QUALITY CONTROL CRITERIA/REPAIRS
See the Education section

Fig. 3-1. (continued.)

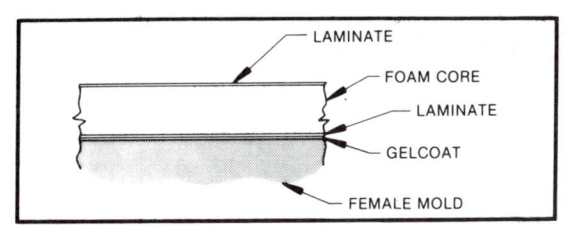

Fig. 3-2. The composite materials are applied to a mold, resulting in a sandwich. (courtesy Stoddard-Hamilton Aircraft, Inc.)

an aircraft with a method you are interested in, then get an invite to his shop. You can learn more in an hour or two with a knowledgeable builder than you can by reading for days.

Also look at the aircraft specifications section of this book. If you see something that strikes your fancy, send for the information package from the manufacturer (there is usually a nominal charge for these packages).

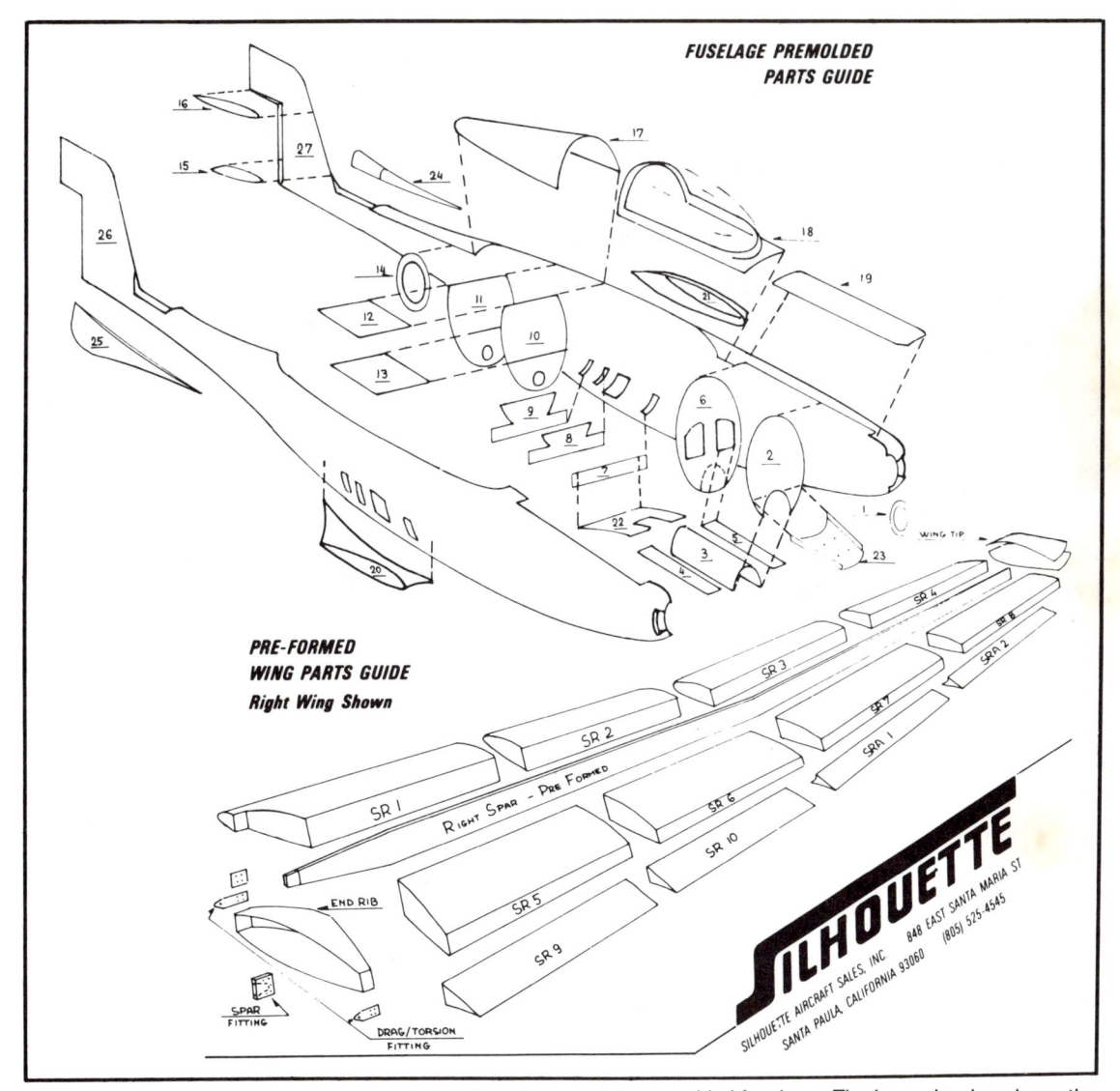

Fig. 3-3. The upper drawing shows the exploded view of a complete premolded fuselage. The lower drawing shows the parts and placements for building a wing using preshaped foam cores. (courtesy Task Research, Inc.)

19

FINAL ASSEMBLY

NOTE: Bulkheads 7, 8 & 9 must be inserted into the fuselage halves before they are joined. Leave them loose. They will be bonded later.

12. Disassemble the fuselage halves by removing the bolts from the rudder hinge brackets and removing all the clecos except the ones holding the rudder nut plate brackets to the right fuselage half.

13. Clean all the drilling chips from the edges of the parts. You should have holes about 6" apart all along the joint overlaps.

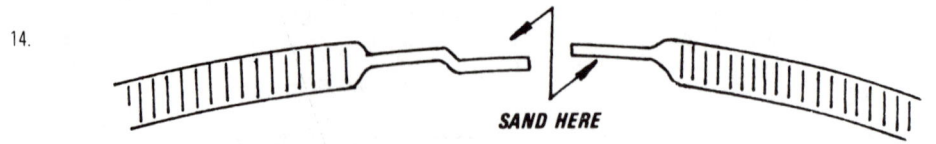

14.

SAND HERE

Sand the joggle mating areas with 40 grit sand paper at all *contact* areas including the mating surfaces of the fin spar and the areas where the fin ribs mate. When completed, dust them off and don't touch them anymore.

15. Plug the rudder hinge bracket nut plates with clay so the threads won't fill with epoxy, and sand the aluminum surface to be bonded with 180 grit sandpaper.

16. Bond and rivet brackets to the right fuselage half.

17. Mix a quantity of resin and cotton flox to bond the right and left halves together. It will take about ½ cup. Add flox until the mixture is still wet, but not runny. Use a spatula and build a small ridge along the underlying joggle like this:

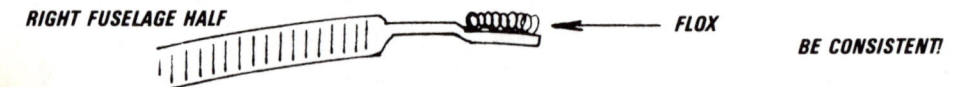

RIGHT FUSELAGE HALF ← **FLOX**

BE CONSISTENT!

Start at the vertical fin and put enough on the mating surface of the ribs to make a bond when assembled.

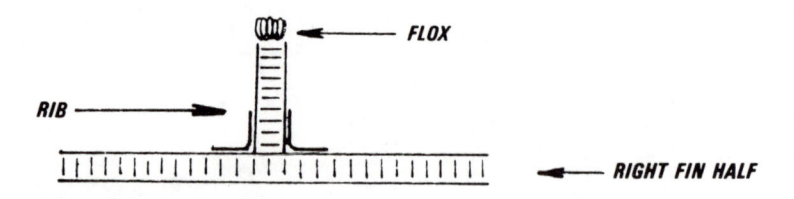

FLOX

RIB →

← **RIGHT FIN HALF**

The closed and finished joint will look like this:

← **LEFT FIN HALF**

"FLOX" SQUEEZE OUT

RIB →

← **RIGHT FIN HALF**

Silhouette ASSEMBLY MANUAL
(Sample Page)

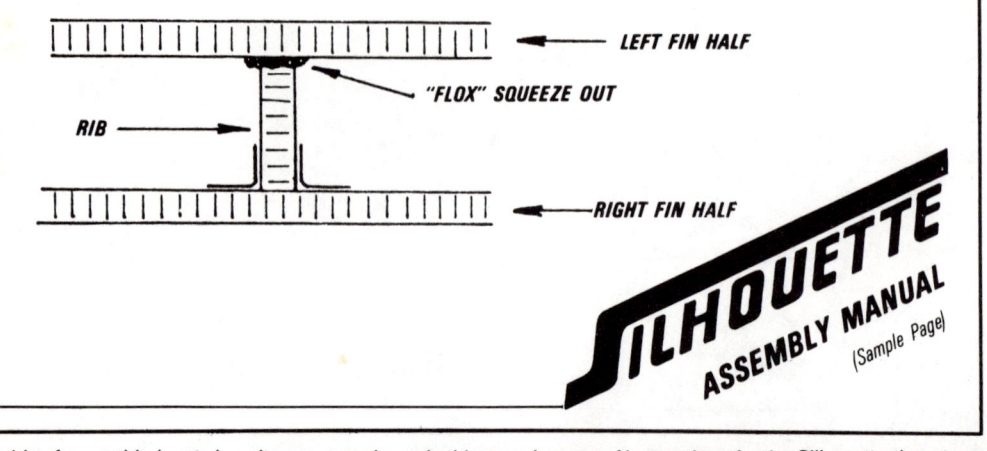

Fig. 3-4. Assembly of premolded parts is quite easy, as shown in this sample page of instructions for the Silhouette. (courtesy Task Research, Inc.)

Chapter 4

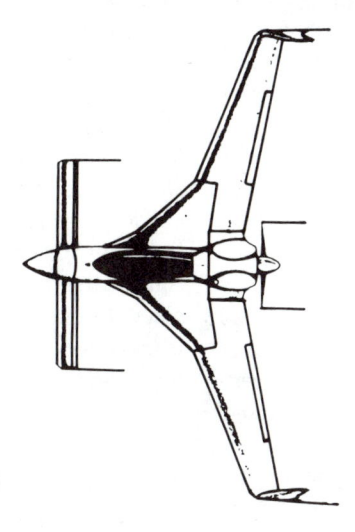

Certified Airplane Engines

The certificated engine is a standard airplane engine that is used in normally certificated aircraft. Examples of these engines are the Teledyne Continental and AVCO Lycoming flat four air-cooled powerplants, the "flat four" referring to the general configuration of the four-cylinder opposed airplane engine.

The use of a certified aircraft engine in a homebuilt airplane results in reduced flight restrictions from the FAA during the initial flight testing phase.

AVAILABILITY

Engines may be purchased new or used from manufacturers, individuals, FBOs, or rebuilders. You will see all these sources listed in the classifieds such as found in *Trade-A-Plane*.

The best engines are new or factory-remanufactured. The factory-remanufactured engine has been completely disassembled, checked for tolerances, needed parts replaced, and reassembled by the factory. It starts a new life with zero time and a clean logbook.

USED ENGINES

Here are a few definitions to help you understand engines better:

Remanufacture: The disassembly, repair, alteration, and inspection of an engine. It includes bringing all specifications back to new limits. A factory-remanufactured engine comes with new logs and zero time.

New limits: The dimensions/specifications used when constructing a new engine. These parts will normally reach TBO (time between overhaul) with no further attention, save for routine maintenance.

Overhaul: The disassembly, inspection, cleaning, repair, and reassembly of the engine. The work may be done to new limits or to "service limits."

Service limits: The dimensions/specifications below which use is forbidden. Many used parts will fit into this category; however, they may not last the full TBO, as they are already partially worn.

Top overhaul: The rebuilding of the head assemblies, but not of the entire engine. In other words, the case of the engine is not split; only the cylinders are pulled. Top overhauls are used to bring oil burning and/or low compression engines within specifications. It is a method of stretching the life of an otherwise sound engine. A top overhaul is not necessarily an indication of a poor engine. Its need may have been brought on by such things as pilot abuse, lack of care, lack of use of the engine, or plain abuse (i.e., hard climbs and fast letdowns). An interesting note: The term "top overhaul" does not indicate the extent of the rebuild job (number of cylinders rebuilt or the completeness of the job).

TBO (Time between overhaul): The manufacturer's recommended maximum engine life. It has no legal bearing on airplanes not used in commercial service; it's only an indicator. Many well-cared-for engines last hundreds of hours beyond TBO—*but not all*.

Nitriding: A method of hardening cylinder barrels and crankshafts. The purpose is to reduce wear, thereby extending the useful life of the part.

Chrome plating: Used to bring the internal dimensions of the cylinders back to specifications. It produces a hard, machinable, and long-lasting surface. There is one major drawback of chrome plating—longer break-in times. However, an advantage of the chrome plating is its resistance to destructive oxidation within combustion chambers (important if the engine is unused for extended periods of time).

Magnaflux/Magnaglow: Terms associated with methods of detecting invisible defects in ferrous metals (i.e., cracks). Parts normally Magnafluxed/Magnaglowed are crankshafts, camshafts, piston pins, rocker arms, etc.

The next best thing, after a factory-remanufactured engine, is to purchase an engine that has been overhauled by a reliable rebuilder. This engine, although not starting life as new, will probably give long and reliable service. It will be considerably lower in cost than a remanufactured engine.

The last choice is to purchase a used engine. This is the least expensive route to take—at the initial point. However, in the long run, purchasing someone else's troubles may be more expensive than cost-effective. Before you purchase any used engine, have a trusted AP or IA check it out for you. You may have to pay him a few dollars for his time, but you could save thousands by making this small investment.

One last point: Don't purchase *any* used engine that has less than one third of its TBO (time before overhaul) remaining, or an engine that has not been kept in compliance with any applicable ADs (Airworthiness Directives).

Many airplane engine ads proudly state the hours on the engine (i.e., 1576 SMOH). Basically this means that there have been 1576 hours of use since the engine was overhauled. *Not* stated is how it was used, how completely it was overhauled, or what kind of flying the hours represent. There are few standards.

TIME VS VALUE

The time (hours) since new or overhaul is an important factor when placing a value on an engine. The recommended TBO, less the hours currently on the engine, is the time remaining. This is the span you will have to live in.

Three basic terms are used when referring to time on an airplane engine:

- ☐ Low time—First 1/3 of TBO.
- ☐ Mid time—Second 1/3 of TBO.
- ☐ High time—Last 1/3 of TBO.

Naturally, other variables come into play when referring to TBO: Are the hours on the engine since new, remanufacture, or overhaul? What type of flying has the engine seen? Was it flown on a regular basis? Lastly, what kind of maintenance did the engine get? The logbook should be of some help in resolving any questions about maintenance.

Engines from airplanes that have not been flown on a regular basis—and maintained in a like fashion—will never reach full TBO. Manufacturers refer to regular usage as 20 to 40 hours monthly. However, there are few privately owned airplanes meeting the upper limits of this requirement. Let's face it: Most of us don't have the time or money required for such regular use. This 20 to 40 hours monthly equates to 240 to 480 hours yearly. That's a lot of flying.

When an engine isn't run, acids and moisture in the oil will oxidize (rust) engine components. In addition, lack of lubrication movement will cause the seals to dry out. Left long enough, the engine will seize and no longer be operable.

Beware of the engine that has just a few hours on it since an overhaul. Perhaps something is not right with the overhaul, or it was a very cheap job, just to make the engine more salable. Have your mechanic check it out.

Engines are expensive to rebuild/overhaul; even a Continental C-series will cost $2000 to $3000 to have rebuilt. And it is the rule to spend $4000 and up on engines from larger four-place airplanes.

ENGINE CODES

The model number of an aircraft engine will usually describe either the horsepower:

- Continental C-85 (85 hp)
- Continental A-65 (65 hp)

... or the number will indicate the cubic inch displacement of the engine:

- Lycoming O-235 (235 cubic inch displacement).

The O that appears as part of many airplane engine model numbers merely indicates that the engine is horizontally *opposed* in configuration. All the engines that you will find in common usage today are horizontally opposed.

Suffix codes describe individual models of engines and their specific information (magnetos, timing, balancing, etc). Example: O-235-L2C opposed cylinder engine of 235 cubic inch displacement, L2C model (which in this case is the 118-hp version).

Cylinder Color Codes

Cylinder color codes are applied via paint, or by banding part of the lower cylinder.

- Orange indicates a chrome-plated cylinder barrel.
- Blue indicates a nitrided cylinder barrel.
- Green means that the cylinder barrel is .010 oversize.
- Yellow is used for .020 oversize.

Spark Plug Color Codes

Spark plug color codes identify the reach length of the required plugs. The color will be seen in the fin area of the cylinder between the plug and the rocker box.

- Grey or unpainted indicates short-reach plugs.
- Yellow indicates long-reach plugs.

ENGINE APPLICATIONS/SPECIFICATIONS

There are only two makes of certificated airplane engines commonly available to the homebuilder. These are Teledyne Continental and AVCO Lycoming.

Teledyne Continental

Of the smaller horsepower engines, there are no doubt more Continentals available than Lycomings. The Continental O-200 engines are plentiful, and reasonably priced. None are currently produced.

Models within the early A and C series are grouped according to rated power, the groups being 65, 75, 85, and 90. These four model numbers are prefixed by the series designation A or C.

Models in the C75 group differ from the corresponding dash-numbered models in the C85 group (Fig. 4-1) only in calibration of the carburetor installed. C90 models differ from C75 and C85 models in the design of several major parts and accessories, including the crankshaft, camshaft, crankshaft gear (in -8 models only), carburetor, oil sump, connecting rods, pistons and valve springs. The O-200 (Fig. 4-2) in turn, differs from the C series in the design of its crankcase, camshaft, crankcase cover, carburetor, and oil sump. It differs further in that shielded ignition is standard equipment. Additionally, the C90 and O-200 engines have longer piston stroke and higher compression ratings than their predecessors.

Fig. 4-1. A popular engine, this C85 has been out of production for many years. However, used and rebuildable examples can be found. (courtesy Teledyne Continental)

Following the series letter and power designation and separated from them by a dash is a figure, and, in some instances, a suffix letter or two. This complete model number denotes the installation of certain parts or equipment designed to adapt the basic engine to various classes of aircraft. Those dash numbers and suffix letters that have been used to identify production models built to date are as follows:

- ☐ -8: No provision for starter or generator.
- ☐ -12: Starter, generator, and associated parts installed.
- ☐ -14: Lord engine mount bushings installed; otherwise like the -12.
- ☐ -16: Vacuum pump adapter; otherwise like the -12 models.
- ☐ F: Flange-type crankshaft installed (replaces tapered shaft).
- ☐ H: Crankcase and crankshaft adapted to feed oil to hydraulic controllable pitch propeller.

Conversion of C75 and C85 models to corresponding dash-numbered C90 models is not approved due to the nature and extent of parts differences and to the possibility of unsatisfactory results. It is not possible to convert a -8 model to

24

a -12 model, because the -8 crankcase is not adaptable to the -12 crankcase cover in several respects. Conversion of -12 models to -8 models is not approved for similar reasons. Neither the -8 or -12 models can be converted to -14 in the field because of the special machining required for Lord mount bushings.

Conversion of the C75 models to corresponding dash-numbered C85 models may be accomplished in accordance with instructions contained in Continental Service Bulletins on the subject. Installation of a flange crankshaft in place of a tapered shaft is considered merely a crankshaft replacement and does not require factory approval or special instructions. However, engine indentification plates bearing model or dash numbers other than those originally assigned cannot be issued unless an application for conversion approval has been submitted and approved by the Continental Factory Service Department.

The following list includes model numbers, hp, takeoff rpm, recommended TBO, and displacement for Teledyne Continental aircraft engines:

Model	hp	T/O rpm	TBO	Displacement
A65	65	2300	1800	171 cu. in.
A75	75	2600	1800	171 cu. in.
C75	75	2275	1800	188 cu. in.
C85	85	2575	1800	188 cu. in.
C90-8F	90	2625	1800	201 cu. in.
C90-12F	90	2625	1800	201 cu. in.
C90-14F	90	2625	1800	201 cu. in.
C90-16F	90	2625	1800	201 cu. in.
C145-2	145	2700	1800	301 cu. in.
O-200A	100	2750	1800	201 cu. in.
O-300A	145	2700	1800	301 cu. in.

Fig. 4-2. The O-200 engine powered the Cessna 150s, and, although out of production, is very available. (courtesy Teledyne Continental)

Fig. 4-3. O-235 engine. Note the starter at the lower front. (courtesy AVCO Lycoming)

Continental engines weigh 157.8 pounds, for the four cylinder jobs, to 268 pounds for the six-cylinder engines, depending on the dash number. Additional weight is imposed by the different equipment installed (i.e., starter 15.5 lbs., generator 10.21 lbs.).

The various applications of certificated Continental engines utilized in current composite airplanes are:

- [ ] VariEze: C-90/O-200 engines
- [ ] Long-EZ: O-200 engines

AVCO Lycoming

There are four models of the AVCO Lycoming airplane engines that concern the homebuilder. These are the O-235 (Fig. 4-3), O-290, O-320 (Fig. 4-4), and the O-360. All are four-cylinder. Their general appearance is the same as Teledyne Continental engines (horizontally opposed).

The following list includes model numbers, hp, takeoff rpm, recommended TBO, and displacement for AVCO Lycoming aircraft engines:

Model	hp	T/O rpm	TBO	Displacement
O-235-C1B	115	2800	2000	233 cu. in.
O-235-C2	115	2800	2000	233 cu. in.
O-235-L2C	118	2800	2000	233 cu. in.
O-290-D	130	2800	2000	289 cu. in.
O-290-D2	140	2800	1500	289 cu. in.

Model	hp	T/O rpm	TBO	Displacement
O-320-A1A	150	2700	1200	319.8 cu. in.
O-320-A2A	150	2700	1200	319.8 cu. in.
O-320-A2B	150	2700	1200	319.8 cu. in.
O-320-B2A	160	2700	1200	319.8 cu. in.
O-360-A1A	180	2700	1200	361 cu. in.
O-360-A1D	180	2700	1200	361 cu. in.

The various applications of certificated AVCO Lycoming engines utilized in current composite airplanes are:

☐ VariEze: O-235 engines

☐ Long-EZ: O-235 engines
O-290 engines
O-320 engines

☐ Glasair: O-320 engines
O-360 engines

A FEW OTHER WORDS

There have been some valve modifications made to Continental engines that you should watch for. These changes are to relieve the valve erosion problems encountered when using 100 LL avgas:

Fig. 4-4. The O-320 engine can be found in great numbers, as they have powered many late airplanes including the Cessna 172 series. (courtesy AVCO Lycoming)

- A-65/A-75: Replace the intake valves with part #639661, and the exhaust valves with part #639662.
- C-75/C-85/C-90/C-145/O-200/O-300: Replace the intake valves with part #641792, and intake valve seats with part #641793.

There are similar modifications available for the AVCO Lycoming engines.

O-320 and O-360 series engines may have the TBO extended to 2000 hours by installing 1/2 inch exhaust valves.

These modifications will be noted in the engine log. Check for them.

Fine wire spark plugs, if approved for your engine, are claimed to be less prone to lead fouling.

STORAGE OF ENGINES

The following information is provided courtesy of Teledyne Continental. It applies to all aircraft engines, and shows to what extent you must go to care for your engine investment. If you purchase an engine prior to the completion of your airframe, you should consider the following information. It could save you much money.

General

Engines in aircraft that are flown only occasionally tend to exhibit cylinder wall corrosion more than engines in aircraft that are flown frequently.

Of particular concern are new engines or engines with new or freshly honed cylinders after a top or major overhaul.

In areas of high humidity, there have been instances where corrosion has been found in cylinders after an inactive period of only a few days. When cylinders have been operated for approximately 50 hours, the varnish deposited on the cylinder walls offers some protection against corrosion.

Obviously, even then, proper steps must be taken on engines used infrequently to lessen the possibility of corrosion. This is especially true if the aircraft is based near the seacoast, or in areas of high humidity, and flown less than once a week.

In all geographical areas, the best method of preventing corrosion of the cylinders and other internal parts of the engine is to fly the aircraft at least once a week long enough to reach normal operating temperatures, which will vaporize moisture and other by products of combustion.

Aircraft engine storage recommendations are broken down into the following categories:

- Flyable storage (7 to 30 days).
- Temporary storage (up to 90 days).
- Indefinite storage.

Flyable Storage (7 to 30 days)

a. Service aircraft per normal airframe manufacturer's instructions.

b. Each seven days during flyable storage, the propeller should be rotated by hand without running the engine. Rotate the engine six revolutions, stop the propeller 45 to 90 degrees from the original position. For maximum safety, accomplish engine rotation as follows:

1. Verify magneto switches are OFF.
2. Throttle position, CLOSED.
3. Mixture control, IDLE CUT-OFF.
4. Set brakes and block aircraft wheels.
5. Leave aircraft tiedowns installed and verify that the cabin door latch is open.
6. Do not stand within the arc of the propeller.

c. If at the end of thirty (30) days the aircraft is not removed from storage, the aircraft should be flown for thirty (30) minutes, reaching, but not exceeding, normal oil and cylinder temperatures. If the aircraft cannot be flown it should be represerved in accordance with B Temporary storage or C Indefinite storage.

Temporary Storage (Up to 90 days)

a. Preparation for Storage

1. Remove the top spark plug and spray atomized preservative oil, (Lubrication Oil-Contact and Volatile Corrosion-Inhibited, MIL-L-46002, Grade 1) at room temperature, through upper spark plug hole of each cylinder with the piston in approximately the bottom dead center position. Rotate

crankshaft as each pair of opposite cylinders is sprayed. Stop crankshaft with no piston at top dead center.

NOTE

Listed below are some approved preservative oils recommended for use in Teledyne Continental engines for temporary and indefinite storage:

MIL-L-46002, Grade 1 Oils:

- ☐ NOX RUST VCI-105
  Daubert Chemical Co.
  4700 S. Central Ave.
  Chicago, IL
- ☐ TECTYL 859A
  Ashland Oil, Inc.
  1401 Winchester Ave.
  Ashland, KY

2. Re-spray each cylinder without rotating the crankshaft. To thoroughly cover all surfaces of the cylinder interior, move the nozzle or spray gun from the top to the bottom of the cylinder.

3. Re-install spark plugs.

4. Apply preservative to engine interior by spraying the above specified oil (approximately two ounces) through the oil filler tube.

5. Seal all engine openings exposed to the atmosphere using suitable plugs, or moisture-resistant tape, and attach red streamers at each point.

6. Engines with propellers installed that are preserved for storage in accordance with this section should have a tag affixed to the propeller in a conspicuous place with the following notation on the tag: "DO NOT TURN PROPELLER—ENGINE PRESERVED."

b. Removal From Storage

1. Remove seals, tape, paper, and streamers from all openings.

2. With bottom spark plugs removed from the cylinders, hand-turn propeller several revolutions to clear excess preservative oil, then re-install spark plugs.

3. Conduct normal start-up procedure.

4. Give the aircraft a thorough cleaning and visual inspection. A test flight is recommended.

Indefinite Storage

a. Preparation for storage

1. Drain the engine oil and refill with MIL-C-6529 type.

2. Start engine and run until normal oil and cylinder head temperatures are reached. The preferred method would be to fly the aircraft for thirty minutes. Allow engine to cool to ambient temperature. Accomplish steps a. in Flyable Storage and a. 1 through 6 in Temporary Storage.

NOTE

MIL-C-6529 type 2 may be formulated by thoroughly mixing one part compound MIL-C-6529 type 1 (ESSO Rust-Ban 628, Cosmoline No. 1223 or equivalent) with three parts new lubricating oil of the grade recommended for service (all at room temperature).

3. Apply preservative to engine interior by spraying MIL-L-46002, grade 1 oil (approximately two ounces) through the oil filler tube.

b. Install dehydrator plugs MS27215-2, in each of the top spark plug holes, making sure that each plug is blue in color when installed. Protect and support the spark plug leads with AN-4060 protectors.

c. If the carburetor is removed from the engine, place a bag of desiccant in the throat of the carburetor air adaptor. Seal the adaptor with moisture-resistant paper and tape, or a cover plate.

d. Place a bag of desiccant in the exhaust pipes and seal the openings with moisture-resistant tape.

e. Seal the cold air inlet to the heater muff with moisture-resistant tape to exclude moisture and foreign objects.

f. Seal the engine breather by inserting a dehydrator MS27215-2 plug in the breather hose and clamping in place.

g. Attach a red streamer to each place on the engine where bags of desiccant are placed. Either attach red streamers outside the sealed area with tape, or to the inside of the sealed area with safety wire to prevent wicking of moisture into the sealed area.

h. Engines with propellers installed that are preserved for storage in accordance with this section should have each propeller tagged in a conspicuous place with the following notation on the tag: "DO NOT TURN PROPELLER—ENGINE PRESERVED."

As an alternative method of indefinite storage, the aircraft may be serviced in accordance with the procedures under Temporary storage providing the airplane is run-up at maximum intervals of 90 days and then re-serviced per the temporary storage requirements.

Procedures necessary for returning an aircraft to service are as follows:

a. Remove the cylinder dehydrator plugs and all paper, tape, desiccant bags, and streamers used to preserve the engine.

b. Drain the corrosion preventative mixture and reservice with recommended lubricating oil.

WARNING

When returning the aircraft to service, do not use the corrosion preventive oil referenced in Indefinite Storage a. 1. for more than 25 hours.

c. With bottom plugs removed, rotate propeller to clear excess preservative oil from cylinders.

d. Re-install the spark plugs and rotate the propeller by hand through the compression strokes of all the cylinders to check for possible liquid lock. Start the engine in the normal manner.

e. Give the aircraft a thorough cleaning, visual inspection, and test flight per airframe manufacturer's instructions.

Aircraft stored in accordance with the indefinite storage procedures should be inspected per the following instructions:

a. Aircraft prepared for indefinite storage should have the cylinder dehydrator plugs visually inspected every 30 days. The plugs should be changed as soon as their color indicates unsafe conditions of storage. If the dehydrator plugs have changed color in one-half or more of the cylinders, all desiccant material on the engine should be replaced.

b. The cylinder bores of all engines prepared for indefinite storage should be resprayed with corrosion preventive mixture every 6 months, or more frequently if bore inspection indicates corrosion has started earlier than 6 months. Replace all desiccant and dehydrator plugs. Before spraying, the engine should be inspected for corrosion as follows: Inspect the interior of at least one cylinder on each engine through the spark plug hole. If cylinder shows start of rust, spray cylinder corrosion preventive oil and turn prop 6 times, then respray all cylinders. Remove at least one rocker box cover from each engine and inspect the valve mechanism.

Chapter 5

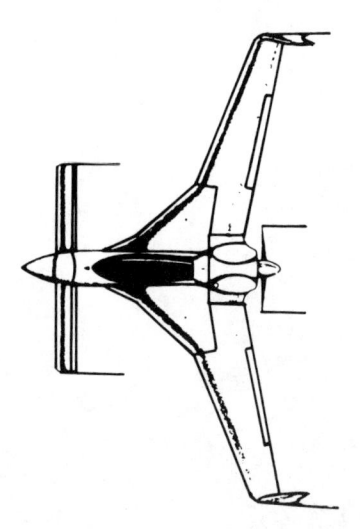

Non-Certified Airplane Engines

There are many non-certified airplane engines available. The examples I will give here include the "airplane type" (air- cooled horizontal opposed), and the new rotary-style engines. Although the sky is the limit, and various liquid-cooled or oddly shaped engines can be used, the air-cooled is the most common.

AMI ENGINES

The AeroMotion Twin (Fig. 5-1) as described in AMI's literature:

The design problem was obvious: Build a lightweight, low-cost four-stroke engine in the 50-55 hp range for ARVs, two-place ultralights, and VW-powered aircraft. To do so, AeroMotion adhered to strict guidelines:

It had to be an aircraft engine, not a rehashed design that was originally designed for another purpose.

Dual ignition with aircraft magnetos was a must.

Simplicity was paramount; there was no need for complexity.

Low weight, under 100 pounds, was the goal.

It had to have enough displacement to give its horsepower at less than 3,200 rpm to eliminate reduction units.

Existing, proven components must be used where possible to keep costs down and reduce the unknowns.

In other words, the design was to be a dirt-simple engine that did the job with a minimum of muss and fuss.

The final product, the AeroMotion Twin, is just what it was supposed to be. At 113 pounds, it cranks out 50-plus horses at 3100 rpm. It's a little old-fashioned in that it breaks no technological barriers; an airplane engine is the wrong place to try something new and exotic. It's an opposed flat-twin that uses a maximum of off-the-shelf automotive V-8 parts, such as pistons, valves, all bearings, etc.

These parts were designed for much harder use in autos and give the AeroMotion Twin great reliability. Another part of that reliability is the pair of Slick Aircraft magnetos that fire the twin plugs. The prototype engine has been running since April

Fig. 5-1. AMI Twin. (courtesy AeroMotion, Inc.)

of 1982 and hasn't so much as hiccupped, according to AeroMotion.

For further information contact:

AeroMotion, Inc.
1224 W. South Park Ave.
Oshkosh, WI 54901

Specifications:

Output:	52 hp.
Max rpm:	3100.
Displacement:	100 cu.in.
Crankshaft:	Forged.
Fuel:	80-87, 100 LL, or auto low lead.
Fuel consump:	2-3 gph.
Weight:	113 lbs.
TBO:	1500 hr. estimate.
Auto parts:	Valves.
	Valve seats and guides.
	Rocker arms.
	Pistons and rings.
	Connecting rods.
	All bearings and seals.
	Timing gears.
	Oil pump.

LIMBACH FLUGMOTOREN ENGINES

Limbach Flugmotoren is a privately owned business located near Bonn, West Germany. The firm manufactures German government-certified engines for most of the motorgliders made throughout the world. The company has a long history of supplying high quality engines to the light/sport aircraft industry.

Fig. 5-2. Limbach engine, which can be configured as SL1700D (65 hp). (courtesy Limbach Aircraft Engines)

At the present time, nearly all of the Limbach Flugmotoren's production is sold directly to airframe manufacturers. A few individual engines have been imported to this country for use with homebuilt aircraft. Special sales of this type can be arranged through:

>Limbach Aircraft Engines
>Box 1201
>Tulsa, OK 74101

This firm is developing a complete sales, parts, and service organization in Tulsa.

The product line of the Limbach Flugmotoren includes many four-cylinder aircraft engines in the 60 to 90-hp range. Most of the engines currently being produced are of the 1700cc (68 hp) and 2000cc (80 hp) size, and vary otherwise according to the accessories supplied. A project to develop a larger engine has recently been started. The new engine will be in the 2500cc to 2700cc range, and produce at least 100 hp.

Limbach Flugmotoren engines are original in design, and are not related to any other German engines (Figs. 5-2, 5-3).

Specifications:

Model SL 1700 EA 1
Output: 60 hp.
Max rpm: 3550.
Displacement: 102.52 cu.in.
Fuel: 96 octane.
Fuel consump: 3-4 gph.
Weight: 149.6 lbs.

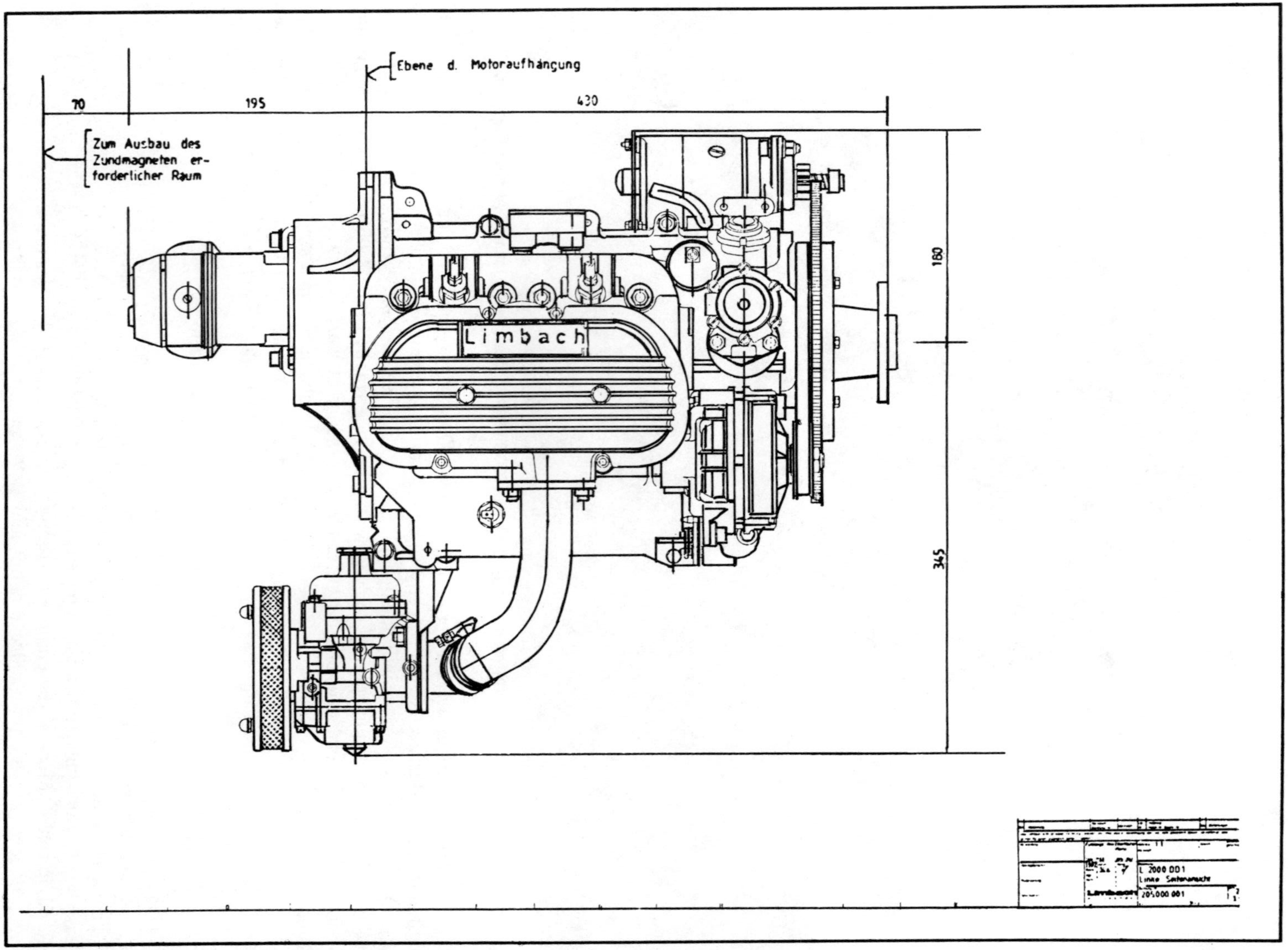

Fig. 5-3. Limbach L2000 DD "uncertified" engine. Notice the location of the starter, flywheel, and alternator. (courtesy Limbach Aircraft Engines)

Model L 2000 EB1-A,B,C,DD

Output:	80 hp.
Max rpm:	3400.
Displacement:	121.68 cu.in.
Fuel:	96 octane.
Fuel consump:	4-5 gph.
Weight:	159.5 lbs.

VW CONVERSIONS

VW engine conversions can be obtained in two ways—outright purchase of a factory conversion, or purchase and installation of various parts on an existing block.

FACTORY H.A.P.I. ENGINES

The people at the H.A.P.I. engine company pride themselves on the fact that all their engines are built from the highest quality materials by skilled technicians, using the best equipment and procedures. Each factory-built H.A.P.I. engine is made to order, with the options the customer desires.

H.A.P.I. engines use only forged crankshafts. H.A.P.I. states: "Only a forged crankshaft will withstand the stresses it encounters in aircraft applications" (Figs. 5-4, 5-5).

Specifications:

Model 50-E

Output:	55 hp.
Max rpm:	3200.
Fuel:	Auto gas.
Fuel consump:	3-4 gph.
Weight:	145 lbs.
Displacement:	1680cc.

Fig. 5-4. H.A.P.I. 60-2DM dual ignition VW conversion. (courtesy H.A.P.I. Engines, Inc.)

Fig. 5-5. H.A.P.I. Series 75 engine. (courtesy H.A.P.I. Engines, Inc.)

Model 60-E
Output: 60 hp.
Max rpm: 3200.
Fuel: Auto gas.
Fuel consump. 3-4 gph.
Weight: 145 lbs.
Displacement: 1835cc.

-2 engines have alternator and electric starting.
-TC engines are turbocharged.

For further information about H.A.P.I. engines contact:

H.A.P.I. Engines Inc.
Eloy Municipal Airport
RR# 1 Box 1000
Eloy, AZ 85231

INAV (FORMERLY MONNETT) ENGINES

The Monnett VW conversion engines had their start in Formula Vee air racing. Formula Vee racing proved that economically reliable high performance is possible with converted 1600 and 1700cc VW engines.

The Monnett engines are basically stock

engines, of new manufacture. Monnett does not use any oversize cylinders, as they feel that these modifications reduce the overall reliability of the engines (Fig. 5-6).

Specifications:

Horsepower:	Up to 70.
Type:	VW III (1969 and up).
Displacement:	1600cc (96.6 cu.in.).
Compression:	7.7:1.
Weight:	150 lbs. (with alternator).
Fuel:	Auto gas.

For further information about Monnett engines contact:

INAV Ltd.
P.O. Box 2984
Oshkosh, WI 54901

BUILDING A CONVERSION

The following pages are provided courtesy of Rex Taylor, from his book *How to Build a Reliable VW Aero Engine*. I recommend this book to anyone contemplating building/converting his own VW engine. The secret of a beautiful engine is very simple—just put it together right. *Never*—I repeat—*never hurry; check and recheck* each and every step before going to the next one. If you are not 100 percent satisfied that what you have just finished is done as well as it can possibly be done, then you are not finished.

One thing you should always keep in mind above every other consideration—you will bet your life on this engine every time you fly behind it, so take whatever time it takes to do each small job to perfection. You will finish your engine knowing you can depend on it, because you will be able to see each and every part in your mind, secure in the

Fig. 5-6. Monnett Aero Vee VW conversion (courtesy Monnett Experimental Aircraft)

knowledge that you did it right, so therefore your engine will not let you down.

So now it is time to start making your engine decisions. You have the airplane you want to build, or may even have it near completion, and your decision is whether to build it (engine) yourself, or buy an engine from those of us who convert them as a business. There are good and bad points either way you decide to go.

If you go with the factory-converted engine, you will have to spend a good sized chunk of your aircraft cost—usually in one place, at one time. Of course you get an engine that you have not expended hours of labor on, but sometimes the wait with some brands of engines is a year or more before delivery. Even purchasing a ready-to-run conversion, you will still have to do a lot of installing and cowling work, which is really part of the powerplant. The big advantage to buying a factory conversion is that a certain wealth of Volkswagen conversion experience is built into the engine. Some conversions are better done than others, and the more time spent in comparison shopping when purchasing one of these conversions, the more likely you are to get the reliability and performance you are looking for. The professional converter has probably gone through an extensive test-run period, and has engines in use in various aircraft. He has had the input from his customers concerning problem areas, so he has probably worked out most of the bugs in his conversion design.

Another important factor worthy of your consideration is the area of warranty or guarantee that you get from the conversion manufacturer. Most legitimate engine builders will stand behind their engines to some degree should you have trouble with one of them. Should you build your own, you become your own guarantor and will be totally responsible for any problems that may develop.

The question that quite naturally comes to mind is whether or not you, as an inexperienced person in the area of Volkswagen aircraft conversions, can manufacture and assemble an engine that has the same degree of reliability as one you can buy. I think the answer to this question is a definite *yes*, if (and here comes the rub) you have, or can develop, all of the following essentials:

1. A clean, dust-free place to work on your engine.

2. Proper tools to perform each stage of the assembly.

3. Reference material available to guide you through each operation, and assure it is done properly.

4. The machine shop tools or machine shop services available for necessary machining.

5. The patience and study time necessary to understand each step before you start it—then do it, and possibly redo it, until it is right.

If you have these five things going for you, or at least most of them, you can build your own engine. On a cost-saving basis, if you put a reasonable price per hour on your time, it will probably cost you more to build your own engine than to buy it. In terms of pride and accomplishment, you will gain an understanding of your engine, and indeed may well find that working on engines is a very enjoyable pastime. Most of the things you will be exposed to will help you in tinkering with the family automobile.

If, however, you are an impatient person who does not enjoy working to exacting tolerances, or do not like to get your hands oily, or perhaps you are one of those people with a "that's good enough" attitude who is satisfied with less-than-perfect in the jobs that you do accomplish, do yourself a favor and buy an engine! There is only one way to build an engine right, that is to build it the very best it can be, or do not do it at all!

Now let's look at each of these five essentials in order to clarify them:

1. The clean, dust-free area necessary need not be large. A small corner of your garage should be sufficient. I will assume you have some kind of a sturdy table at a comfortable working height to work on, and probably a bench vise. Two very useful tools in the building of a Volkswagen engine are an engine buildup stand, which holds your engine in any position while you work on it, and a plain drip pan such as is used under automobiles to catch dripping oil on garage floors. The pan is used on top of the bench as your spotlessly clean

surface for building up subassemblies and laying out your tools. Incidentally, your tools should be cleaned in solvent dust and dirt-free before using them on your engine. It is a little like a surgeon operating—the work area must be free from contamination by dust and dirt.

Since this process will undoubtably take a few days spare time at the very least, your engine and its parts will require protection from dirt during storage periods. We have found that plastic garbage bags do an excellent job as dust covers capable of enveloping a complete engine on its assembly fixture, and really protecting it from dirt until the next time you can work on it.

2. The tools required to build up your engine will require an investment of possibly $150 if you don't already have them. You will need a 3/8 drive socket set, and a set of box/open end wrenches in metric from 10mm to 17mm sizes. You will also need a ring compressor, screwdrivers, pliers, Allen wrenches, and—most important of all—an accurate torque wrench. If you think you can just guess your way through and avoid buying a torque wrench, forget building an engine at this point; you don't have the right attitude, and you are going to produce a piece of undependable junk. Either condition yourself to do the job totally right, or don't do it at all.

You will require a few more tools such as a propane torch, possibly a gear puller if you have to pull your prop hub for some reason after assembly, a dial indicator, a 2 to 3-inch micrometer, and an ohm meter for timing your magneto.

Liberate some of your wife's baking pans to store your nuts and bolts as they are removed from your engine. Separate them into engine sections, such as head bolts, crankcase bolts, internal engine parts etc.

3. The reference materials needed will hopefully be met entirely by this book [*How to Build a Reliable VW Aero Engine*] in the step-by-step assembly section. There are some other publications available that can add to your reference library such as the Type II Engine Maintenance Manual, available at your Volkswagen dealer. If you are a person who has never had much experience in engines, I suggest looking in the public library for a good book on four-cycle engine theory and operation since basically all four-cycles are very much alike.

4. The machine shop services or machining ability and availability you will find necessary are probably the largest stumbling block for some. There is quite a bit of modification and machine work necessary to build larger engines than 1680cc. With the larger bore sizes the engine crankcase and cylinder heads must be bored out to accept these larger assemblies. While the operations necessary are not too complex or difficult to do, to have them done in a shop that has no previous experience with these operations usually proves very costly because the shop would not have developed holding fixtures and cutting tools to do it in an efficient manner as in a production run. The shop will charge you by the hour usually, and it will be very expensive. There are many firms who do this work on a routine basis for very reasonable prices, and turn out good quality work.

Some of you may have to send your parts off for machining. H.A.P.I. has these machine operations available at competitive prices. Should you be fortunate enough to have your own lathe and milling machine, or access to them, you may well be able to perform these jobs yourself. There is one point to consider here— if you louse it up, you have to throw it away at your own expense, but if the shop louses it up, they must replace it at no cost to you!

5. The last requirement is the most important of all. Those of you with patience, dedication to the task at hand, and the will to finish can probably get by very well without having the most ideal conditions to work in, the best tools to work with, abundance of reference materials for every question, the availability of an expert machinist, or machine tools of your own. If you are not willing to spend the time necessary, or don't have a good track record for finishing projects you start, you probably won't finish this one either. If you don't finish and decide to sell out, you'll probably take a considerable loss on your investment in parts. So be honest with yourself, *about* yourself, before you start.

By now you should have decided what your engine needs are for your particular plane, but you still have some more decisions to make.

We can start building up an engine based on all new parts, or we can use a certain amount of used parts in the process. Either way has advantages and certain disadvantages. New parts of course are more expensive, but are many times a lot easier to get, and you do not have to clean them of years of accumulation of grease and dirt before you use them. Used parts are sometimes obtained very cheap if you are a good scrounger, but you never really know what you have until you have spent hours in teardown, cleaning, and inspecting for damage that will render the part incapable of being rebuilt to aircraft standards. Sometimes you will get lucky and get an engine that is very good inside. Then there are the others that turn out to be junk, in which case you have usually wasted your money as most used parts are sold on an "as is" basis.

If you plan on going the used engine route, don't get in a big rush to buy. Look around the auto wrecking yards in your area and try to find a Volkswagen that was wrecked rather than run out. Look for one that was totaled for front end damage, with the engine and transaxle section undamaged. The transaxle is not important to us, but if it is damaged there is a good chance that the engine case has had severe shocks and stresses too, and is probably also damaged. Many times cases are cracked in accidents, but such cracks are not found until after the engine is disassembled and cleaned.

Your best bet is an engine that can still be started and run in the car. The engines found in the Volkswagen squareback sedans and station wagons that Volkswagen calls "suitcase engines" are a very good source for aircraft. These engines were all built in 1500cc or more displacement, and most have the desirable large oil gallaried case. Best of all, they are not as eagerly sought after as the vertical blower Bug engine, so usually can be bought for less. Some of the later squarebacks had electronic fuel injection and dual port heads. The fuel injection is of no value to us, but the rest of the engine is very suitable for conversion. This engine is an ideal choice if you are going to build a turbocharged engine.

New parts, of course, are much easier to obtain, but even here you will have to know exactly what you want, and be sure you are getting what you want. There are all too many suppliers of replacement parts for Volkswagens now who are turning out parts that *look* like the originals, but are real junk. In order of quality the best parts are usually German, the next best from Japan, and the real junk is from Brazil. This is not a hard-and-fast rule, however, as each of these countries mentioned also turns out some parts of very good quality. The super bargain Volkswagen parts you will see advertised in some catalogs are usually of very shoddy quality even for auto use, and are poison in an aircraft. Price alone is not always a good indicator, but is a fairly reliable one.

At H.A.P.I. we have been offered a set of gaskets for as little as $.93, but we still use the gaskets that cost five times as much, simply because we want the oil on the inside of the engine. The Volkswagen gasket set consists largely of rubber seals, and you want a high-quality, live, heat-resistant, long-life rubber instead of hard inflexible dry junk in the cheap sets offered. If the parts you get will not do the job properly, they are not cheap—even if you get them free.

There is another source of parts in the form of rebuilt parts. There are rebuilt cases, reground crankshafts, rebuilt heads, and even rebuilt connecting rods.

The rebuilt cases can be a good alternative source to a new case if they have been align-bored and reconditioned by a reputable firm who has high standards. A reconditioned case can serve you as well as a new case. The big question here is to know what kind of an outfit you are dealing with. One such company is Rimco, whose cases are equal to new in reliability, but very little less in cost. The sharp person will see right away how Rimco gives you a straight job—true main and cam tunnel align-boring.

Cylinder heads are routinely rebuilt for auto use. They are rebuilt with salvage methods such

40

as helicoil inserts in the plug holes; cracks are repaired by heliarc welding, which is asking for trouble in aircraft. You may find a pair of heads on a used engine with none of the above problems, but be aware that the used engine you buy may have already had less-than-airworthy things done in its past history. New cylinder heads are not too much more in cost than a top-quality rebuilding job, and you can be certain of what you have.

Crankshafts are a different matter, with the used ones being the best ones within certain limits. Volkswagen for many years built their engines with a high-quality forged steel crankshaft that has the bearing journals surface-hardened. These shafts are readily available, and most will be found to have serviceable standard size bearings. Such a shaft, after Magnafluxing, is reground to .010 undersized on the rod and main bearings, and will be far superior to the soft cast steel late-model cranks for use in aircraft. A cast crank when subjected to bending or shock loads is brittle and will break like a piece of glass, whereas a forged crank will bend without breaking. A broken cast crankshaft could easily result in a loss of your propeller. The 1500 crankshaft VW part number 311-353-A is our choice because it has the cross-drilled oil feed holes to the bearings, but it does not have the elongated slots as the later cranks did. The bearing loads imposed by aircraft service cause the elongated slot cranks to wear out of round due to the smaller amount of bearing contact surface.

Rebuilt connecting rods are cheap but should be avoided, as there are too many things that can be wrong with them that may not be apparent on inspection. We use a very expensive rod by comparison, but have never had a connecting rod problem of any kind, so feel that the extra expense is justified.

Most of all the remaining parts in the engine you build will be new except for those stock parts that are used in the non-critical areas on the engine. We will look them over carefully to ensure their airworthiness.

The other parts that are unique to the Volkswagen conversion are the propeller hub and the various castings or machined plates necessary to cover up or alter the function of the various openings in the Volkswagen crankcase.

You will also have to decide at this point on just what kind of conversion you want to build, and how simple or complex you want it to be. Should you want to go full-house with engine accessories and add the turbocharger, your engine will become a much more complex piece of machinery where interrelationships between such things as compression ratio, boost pressure, and engine timing are critical to performance; if not right, it will tend to self-destruct. Conversely, the ultra-simple engine is relatively easy to build, and though it requires equal exacting attention to detail, there are not nearly so many details to keep control over.

Probably, most designers would be very happy if the builder would stick to the simple versions because of their lighter weight in the aircraft. All other things being equal, the light aircraft will always outperform the heavier one. The builder should bear in mind that while the starter, alternator, accessory case is very nice, it does require the use of a voltage regulator, battery, battery cables, ammeter, master switch, master relay, and usually a few more goodies. The total added weight to your aircraft will probably be over 25 pounds. Should you want to add the turbocharger on top of this, figure another 30 pounds for the turbo and its complex manifolding. All of this weight we are talking about here is usually concentrated on the nose of the airplane, and many times requires moving the engine forward to cram all the goodies under the cowl, so that not only do we have a lot more weight up front than the engine designer ever figured on, but we have it farther forward to complicate the problem of center of gravity even further. The sad part of this weight problem is that none of the accessories have added in any way increase the horsepower available, and your airplane will know it when it comes to lifting off the ground.

All of the engines listed in Table 5-1 can be converted by methods described here; however, the engines marked * are the least desirable for conversion:

Though any of the 1500cc or 1600cc engines above will convert to aviation use, the most

Table 5-1. Volkswagen Engines Suitable for Aircraft Conversion.

Type I Engines First built	Numbers	cc	hp
Aug 1966	H-0-000-000	1500	
Aug 1969	B-0-000-000	1600	50
Jul 1971	AD-0-360-022		
	AE-0-000-001	1600	50 Super Beetle
Aug 1971	AD-0-360-025	1600	48 Smog control
	AE-0-558-001	1600	48 Smog control
	AH-0-000-000	1600	48 Smog control
Jul 1972	AD-0-058-001	1600	48
	AE-0-917-063	1600	48
	AH-0-006-900	1600	48
Aug 1972	AD-0-598-002	1600	50
	AE-0-917-264	1600	48
	AF-0-000-802	1600	46
	AH-0-005-901	1600	48
July 1973	AD-0-749-788	1600	50
	AF-0-034-850	1600	46
	AH-0-056-934	1600	48
	AK-0-060-039	1600	48
Type II Engines			
Dec 1963	*8-264-628	1500	
Aug 1964	*8-785-398	1500	
Oct 1964	*8-964-971	1500	
Dec 1964	*816-281	1500	
Aug 1965	*H-0-000-000	1500	
	*L-0-000-000	1500	Smog control
Aug 1966	H-0-183-373	1500	12 volt
Dec 1966	H-0-309-830		
Jul 1967	H-0-761-325		
Aug 1967	B-5-000-001	1600	
Dec 1967	B-5-017-633	1600	
	C-0-000-000	1600	Smog control
Jul 1968	B-5-050-173	1600	
Aug 1968	B-5-050-174	1600	
Dec 1968	B-5-079-928	1600	
Jul 1969	B-5-116-436	1600	
Aug 1969	B-5-116-437	1600	
Dec 1969	B-5-114-597	1600	
Jul 1970	B-5-230-000	1600	
Aug 1970	AE-0-000-001	1600	
Jul 1971	AE-0-529-815	1600	
Type III Engines			
Apr 1961	000-001	1500	
Dec 1961	013-112	1500	
Aug 1962	065-746	1500	
Dec 1962	143-557	1500	
Aug 1963	255-340	1500	
Dec 1963	408-183	1500	
Jul 1964	633-330	1500	
Aug 1964	633-331	1500	
Dec 1964	816-281	1500	
Jul 1965	1-100-000	1500	
Aug 1965	T0-000-001	1600	
Aug 1966	T0-160-001	1600	12 volt
Aug 1967	U0-000-000	1600	Fuel injection
Dec 1972	U0-069-142	1600	Note: All U series are dual port.

desirable engines are the later models of the Transporters that have dual oil relief valves, about 1970 and on, and the bosses on the case for the rear engine mount as used in the bus. These cases were used in the Type II and Type III vehicles.

For detailed step-by-step instructions of the actual conversion process write to:

>H.A.P.I. Engines, Inc.
>Eloy Municipal Airport
>RR#1 Box 1000
>Eloy, AZ 85231

and request the book *How to Build a Reliable VW Aero Engine* by Rex Taylor. There is a nominal charge for this book, but it is well worth it.

THE COMBINATION ALTERNATIVE

The Great Plains Aircraft Supply Co. manufactures a complete line of VW-based engines for sport aviation. These vary from 1600cc to 2180cc in displacement, and are available with a large range of accessories (single/dual ignition, starter, alternator, etc.). They also offer a unique homebuilder program.

GPASC offers the homebuilder a unique "hands-on" engine assembly workshop. In this workshop, the homebuilder will come to GPASC's shop and assemble the engine of his choice. A GPASC staff member will be on hand at all times to provide guidance and assistance to the homebuilder. They will work together hand-in-hand. After the engine is assembled, it will be test-run to ensure its proper operation. The assembly workshop takes about 12 hours to complete.

Engines that are assembled in the workshop program are referred to as Models GP or CP.

By taking advantage of this program, the homebuilder will gain valuable engine maintenance knowledge and realize a savings of $100 to $300, depending on the engine selected (Fig. 5-7).

For more detailed information on this unusual method of engine conversion contact:

>Great Plains Acft. Supply Co., Inc.
>P.O. Box 1481
>Palatine, IL 60078

ROTARY ENGINES

The rotary engine, often called the Wankel after its inventor, is a lightweight, powerful engine, famous for its lack of moving parts. Currently the rotary engines are found in Mazda RX-7 automobiles. Mazda has been producing the rotary engine since 1974.

Duncan Rotary Engines, Inc., is currently producing several versions of the rotary engine for use on airplanes. They state in their literature:

A message to those who have been waiting...

How would you like to carve 90 to 100 pounds of dead weight off of your 1940s technology O-235 aircraft engine, cut its overall size in half (Fig. 5-8), reduce the fuel consumption by 40-50 percent, and while you are at it, increase the power by five horsepower? Or take your nice little Volkswagen conversion engine, leave the weight and fuel flow right where they are, while doubling the horsepower and tripling the TBO at the same time?

All the Duncan engines described here are liquid-cooled and derived from the Mazda RX-7 engine core. They are all available with 35 amp alternators and electric starters. The cooling system consists of a lightweight aluminum water pump and a small radiator. Plumbing is done with aircraft-quality hydraulic hoses. Carburetion is done with the famous Posa super-carb. The ignition system is a "state-of-the-art" dual electronic ignition system, which is considerably lighter than the traditional aviation magneto system.

Rotary engines, due to their internal operations, display extremely low vibration levels. This is because the rotary engines have no crankshaft with pistons going up and down attached to it. There are also no camshaft, valve train, etc. Of course, along with a reduced vibration level comes a low noise level.

A brief exploration of the operation of the

Fig. 5-7. Great Plains VW conversion you can build at the factory. (courtesy Great Plains Aircraft Supply)

rotary engine shows that it is neither four-cycle nor two-cycle. Rather, it takes a little from each. In practice there are five phases in the operation of the rotary engine (Fig. 5-9).

1. *Intake*—As the rotor turns past the intake ports, the fuel/air mixture is drawn into the chamber. There are no valves in the rotary engine.

2. *Compression*—The apex seal (on the rotor) closes the combustion chamber as the rotor turns. Full compression is reached as the side of the rotor reaches the spark plug.

3. *Ignition*—The plug fires, causing the fuel/air mixture to ignite.

4. *Expansion*—The burning fuel/air mixture rapidly expands, exerting force on the face of the rotor, causing it to turn, producing torque.

5. *Exhaust*—As the rotor reaches the exhaust port, the spent combustion gases are expelled.

All of the above makes up the combustion cycle. Looking at the diagram of the engine, you will note there are three surfaces on the rotor. This means there are three complete combustion cycles for each turn of the rotor. The rotor revolves around the eccentric shaft one time for each three turns of the eccentric. The eccentric may be thought of as similar to the crankshaft, and is usually referred to as the *mainshaft*. The rotor and the eccentric shaft are the only two moving parts in the rotary engine. Remember, for a rotor velocity of 3000 rpm, the eccentric shaft will be turning at 9000 rpm.

Due to the high rpm of the mainshaft there is a reduction unit on Duncan engines. This brings the

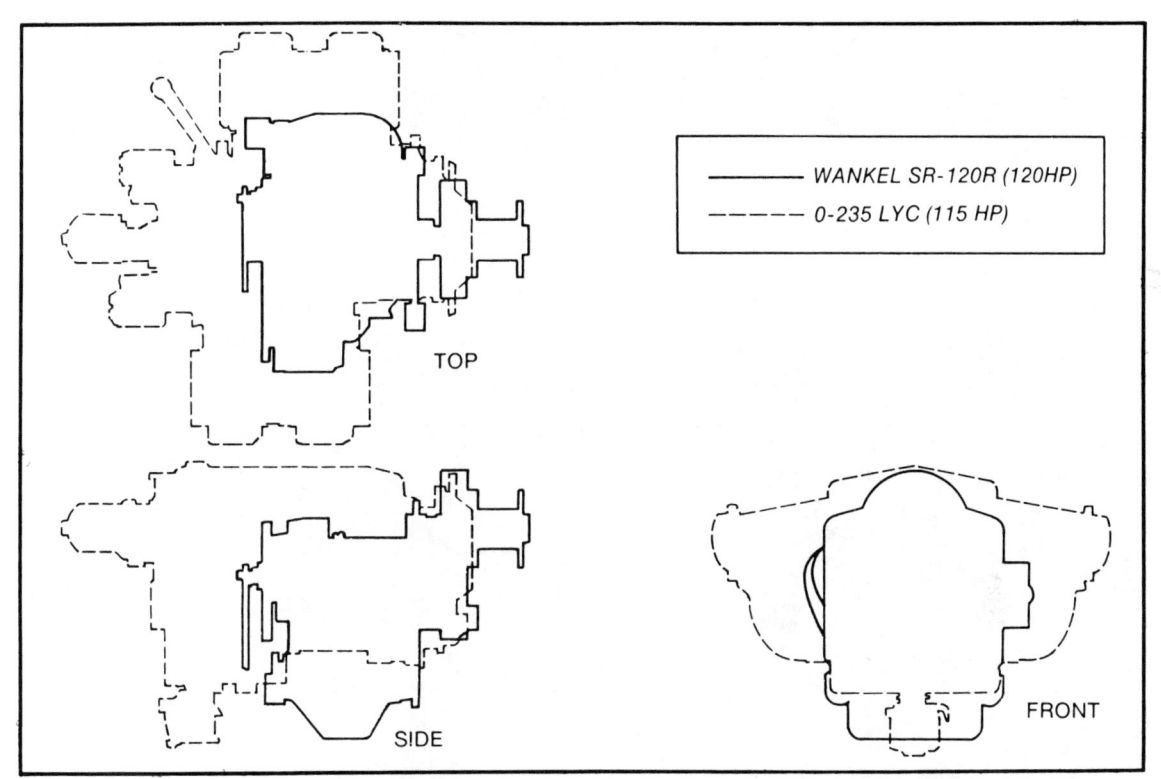

Fig. 5-8. Comparative sizes of the rotary engine and a standard "O" engine. (courtesy Duncan Aviation Engines)

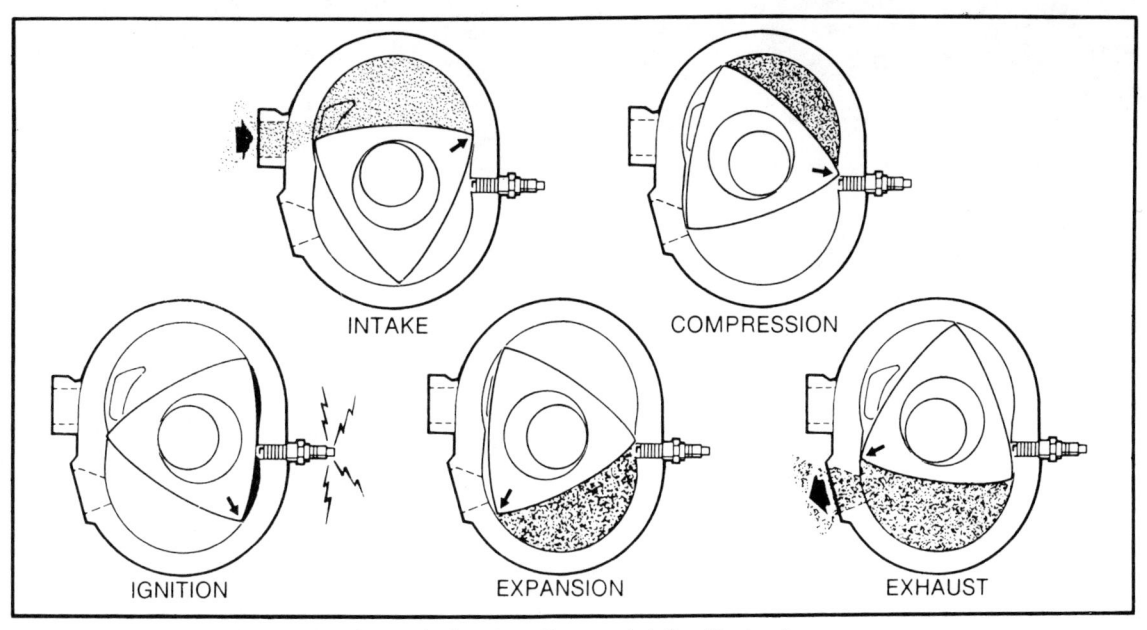

Fig. 5-9. The rotary engine combustion cycle (courtesy Duncan Aviation Engines).

Fig. 5-10. Duncan SR 120. (courtesy Duncan Aviation Engines)

rpm available to the propeller down to a realistic level.

Since there are no valves, rods, lifters, cams, etc., the likelihood of catastrophic failure is considerably reduced from that of a standard piston engine.

Naturally, with a reduction in moving parts there is an increase in the TBO (time between overhaul). Estimated TBO on the Duncan rotary engines is 3000 hours (Figs. 5-10 through 5-12).

Specifications:

Model SR-60D single rotor direct drive
Horsepower:	60 at 4000 rpm.
Max fuel flow:	3.5 gph.
Displacement:	35 cu. in.
Compression ratio:	9.4:1.
Radiator size:	13.75 × 10.5 × 3.62 inches.
Oil cooler size:	8.0 × 8.0 × 3.62 inches.
Fuel type:	Auto (reg or LL), 80 or 100LL avgas.
Weight:	126.3 lbs.
TBO:	3000 hrs.

Model SR-120R single rotor reduction drive
Horsepower:	120 at 9000 rpm.
Max fuel flow:	5 gph.
Displacement:	35 cu. in.
Compression ratio:	9.4:1.
Radiator size:	13.75 × 10.5 × 3.62 inches.
Oil cooler size:	8.0 × 8.0 × 3.62 inches.
Fuel type:	Auto (reg or LL), 80 or 100LL avgas.
Weight:	148.8 lbs.
TBO:	3000 hrs.

Model DR-200R dual rotor reduction drive
Horsepower: 200 at 8100 rpm.
Max fuel flow: 9 gph.
Displacement: 35 cu. in. each rotor.
Compression ratio: 9.4:1.
Radiator size: 13.75 × 10.5 × 3.62 inches.
Oil cooler size: 8.0 × 8.0 × 3.62 inches.
Fuel type: Auto (reg or LL), 80 or 100LL avgas.
Weight: 237.4 lbs.
TBO: 3000 hrs.

Model DR-240R dual rotor reduction drive
Horsepower: 240 at 9000 rpm
Max fuel flow: 10 gph.
Displacement: 35 cu. in. each rotor.
Compression ratio: 9.4:1.
Radiator size: 13.75 × 10.5 × 3.62 inches.
Oil cooler size: 8.0 × 8.0 × 3.62 inches.
Fuel type: Auto (reg or LL), 80 or 100LL avgas.
Weight: 280.3 lbs.
TBO: 3000 hrs.

As an option, all Duncan engines can be operated with constant-speed propellers, or an assortment of standard flange-mounted fixed pitch props.

For further information about these unique engines contact:

Duncan Aviation Engines
Rt 1 Box 256
Comanche, OK 73529

Fig. 5-11. Duncan DR 200. (courtesy Duncan Aviation Engines)

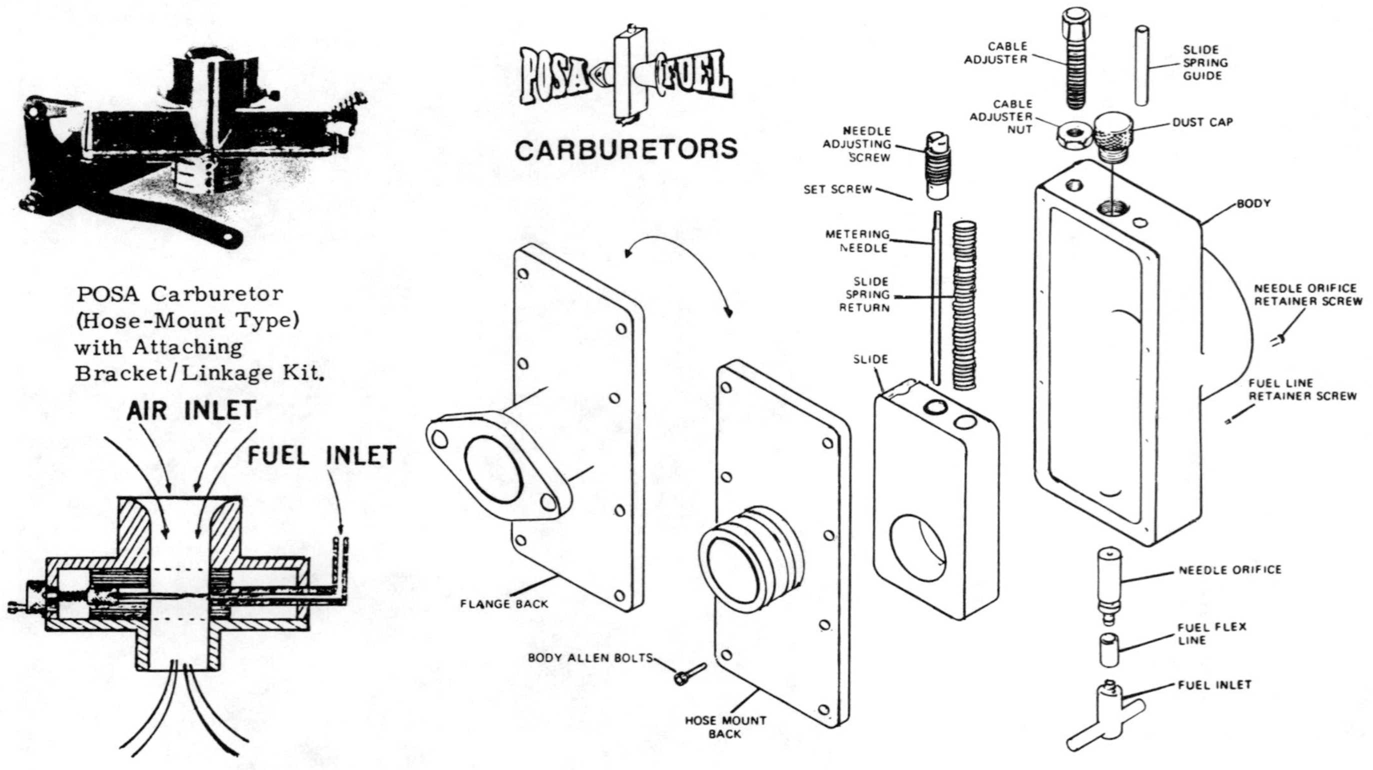

Fig. 5-12. The POSA carburetor, found on most VW conversions and Duncan engines. (courtesy Aircraft Spruce and Specialty Co.)

Chapter 6

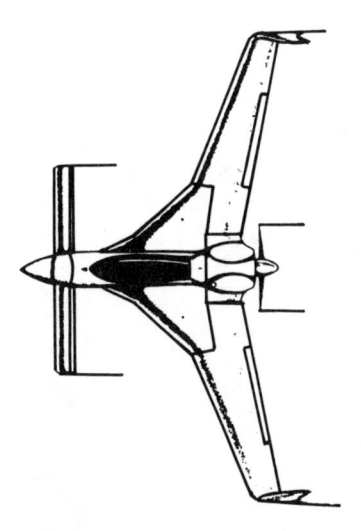

Propellers

The connection between your airplane's engine and the air you fly through is the propeller. Most of the propellers you will encounter in the field of homebuilt airplanes will be constructed of wood (Fig. 6-1).

The following information is reprinted by courtesy of Sensenich Corp., from their bulletin #216:

A. Introduction

Sensenich offers wood propeller models for amateur-built aircraft powered by Lycoming O-290, O-320 and O-360 Engines for use on aircraft built under regulations and standards of FAA publications: AC 20-27B and 20-28B; from plans certified by the National Association of Sport Aircraft Designers (NASAD), or approved by the Experimental Aircraft Association (EAA).

B. Use of Metal Propellers

When the design permits, Type Certificated metal propellers properly selected and matched to the engine and aircraft, and in airworthy condition, may be used. However, reworking of propellers below minimum dimensions of diameter, chord, and thickness specified by the manufacturer or re-pitching above the maximum is a dangerous practice. This is not sanctioned by any manufacturer since it can result in a radical shift of vibration characteristics and lead to early fatigue failure.

It is not economically feasible to develop a metal propeller to Type Certification standards for each of the many designs of amateur-built aircraft since the quantity of any one design is not sufficient to justify the development and tooling costs. Thus, wood propellers offer a practical answer.

C. Why Wood Propellers

Wood has damping properties superior to metal. Wood propellers are made from aircraft-quality birch under quality controlled conditions.

If properly matched to the engine and aircraft, performance in flight approaches that of a metal propeller.

With proper care, frequent inspection, and necessary repair, when in use and storage, wood propellers will give good service. Refer to FAA AC 43:13-1A Chapter 12.

49

Fig. 6-1. Four examples of propellers for amateur-built airplanes. (courtesy Great American Propeller Co.)

D. Propeller Models Available

These wood propellers were originally developed for, and tested on, Thorp T-18 aircraft. They have also been used on other aircraft having similar design, power, and performance characteristics.

It is preferable that the aircraft designer provide the propeller recommendation—Model and pitch—with the plans, so the builder can order accordingly.

If this cannot be done, Sensenich will provide this recommendation providing the customer submits the following data:

1. Model of aircraft
2. Engine make, model, rated hp and rpm
3. Full Throttle Level Flight (FTLF) airspeed of aircraft and density altitude.

Pitch will be in accordance with the data submitted.

Tipping is of epoxy.

Finish is Polyurethane.

Other manufacturers also produce fine propellers. Among these is the Great American Propeller Company. In addition to producing propellers for the standard certificated aircraft engines, they also produce propellers for non-certificated engines found in use on many of the composite airplanes. Also, as noted in the list below, Great American produces propellers for use with pusher design aircraft, such as the VariEze.

The following is reproduced courtesy of Great American Propeller Company:

Aircraft	Engine	Propeller	RPM
Quickie	18.5/22 hp	42 × 28	n/a
Q-2	2100cc VW	56 × 46	3400
Q-200	O-200	62 × 72	2850
Dragonfly	1834cc VW	52 × 42	3600
KR-1	1600cc VW	52 × 42	3500
	1834cc VW	52 × 47	3700
	2100cc VW	52 × 49	3600
KR-2	1834cc VW	52 × 47	3600
	2100cc VW	52 × 48	3600
VariEze	C-90	56 × 66	2900
	O-200	56 × 68	2850
	O-235	58 × 65	2800
Long-EZ	O-200	62 × 58	2850
	O-235	62 × 62	2800
	O-290	62 × 68	2800
	O-320	62 × 72	2800
Glasair	O-320 (150)	68 × 70/72	2800
	O-320 (160)	68 × 72	2800
	O-360 (180)	68 × 74	2800

Each Great American Propeller is 70 percent covered with Dupont Kevlar for maximum protection.

Chapter 7

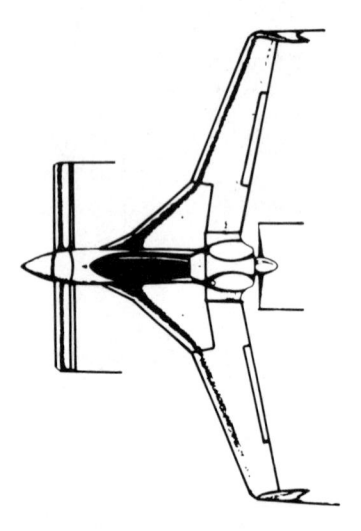

Avionics

Today's modern airway system requires at least a minimum of avionics on board to freely travel where you want to go. Naturally, after you complete your composite airplane, you will want to travel around with it. Your travels may only be local to start with, but it won't be long before you will take advantage of the high cruise speeds offered by these planes, and you will journey hundreds, even thousands of miles.

DEFINITIONS

A-panel—Audio panel. Allows centralized control of all radio equipment.
ADF—Automatic Direction Finder.
CDI—Course Deviation Indicator, a panel-mounted device that gives visual output of the NAV radio.
COMM—VHF transceiver for voice radio communications.
ELT—Emergency Locator Transmitter (required by FARs for all but local flying).
LOC/GS—Localizer/Glideslope. Visual output is via a CDI, with the addition of a horizontal indicator.
Loran—An extremely accurate electronic system of radio location.
MBR—Marker Beacon Receiver.
NAV—VHF navigation receiver for utilizing VORs.
NAV/COMM—Combination of COMM and NAV in one unit.
XPNDR—Transponder (may or may not have altitude encoding).

NEEDS

Now let's examine your flying needs/habits and apply them to the avionics available (Figs. 7-1 through 7-19).

If you are a casual flier and do little or no cross-country flying, then you can get by with a minimum of equipment. This would apply to the pilot who rarely leaves the pattern or practice area.

☐ ELT
☐ COMM

If you do some VFR cross-country flying, you'll

Fig. 7-1. King KX 170B NAV/COMM. (courtesy King Radio Corp.)

need a little more equipment—just for convenience sake, and to utilize those electronic airways our tax money goes for.

- ☐ NAV/COMM
- ☐ XPNDR
- ☐ ELT

Should you be flying all over the country on extensive trips, such as to all the airshows, you might want to add an extra NAV/COMM just as a backup, and an ADF. Although not listed as avionics, many planes have stereo systems installed. If you have the space, you might want to install one. It sure is nice on long trips.

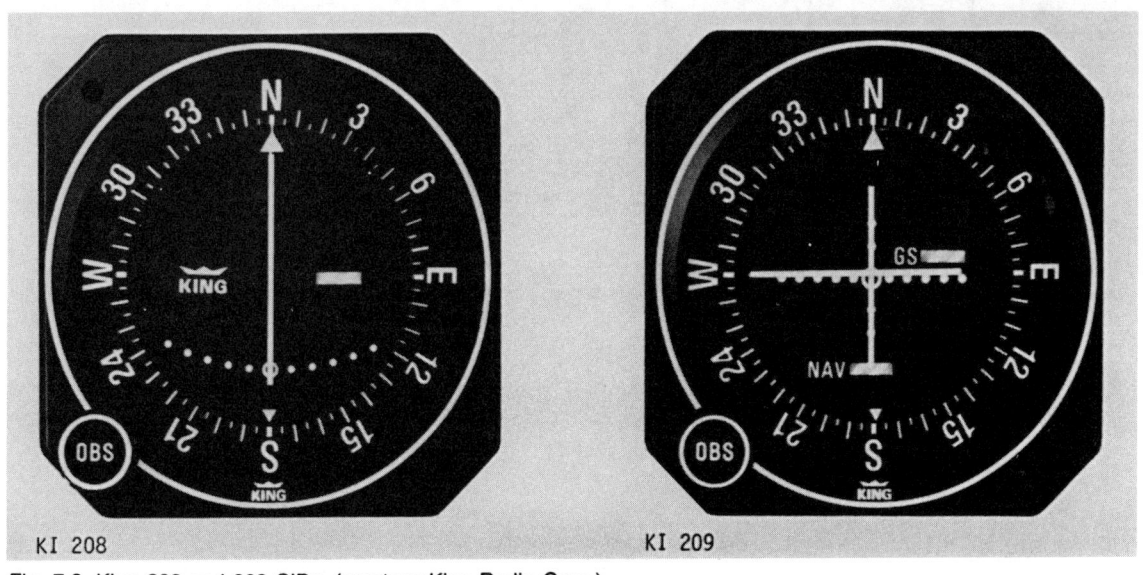

Fig. 7-2. King 208 and 209 CIDs. (courtesy King Radio Corp.)

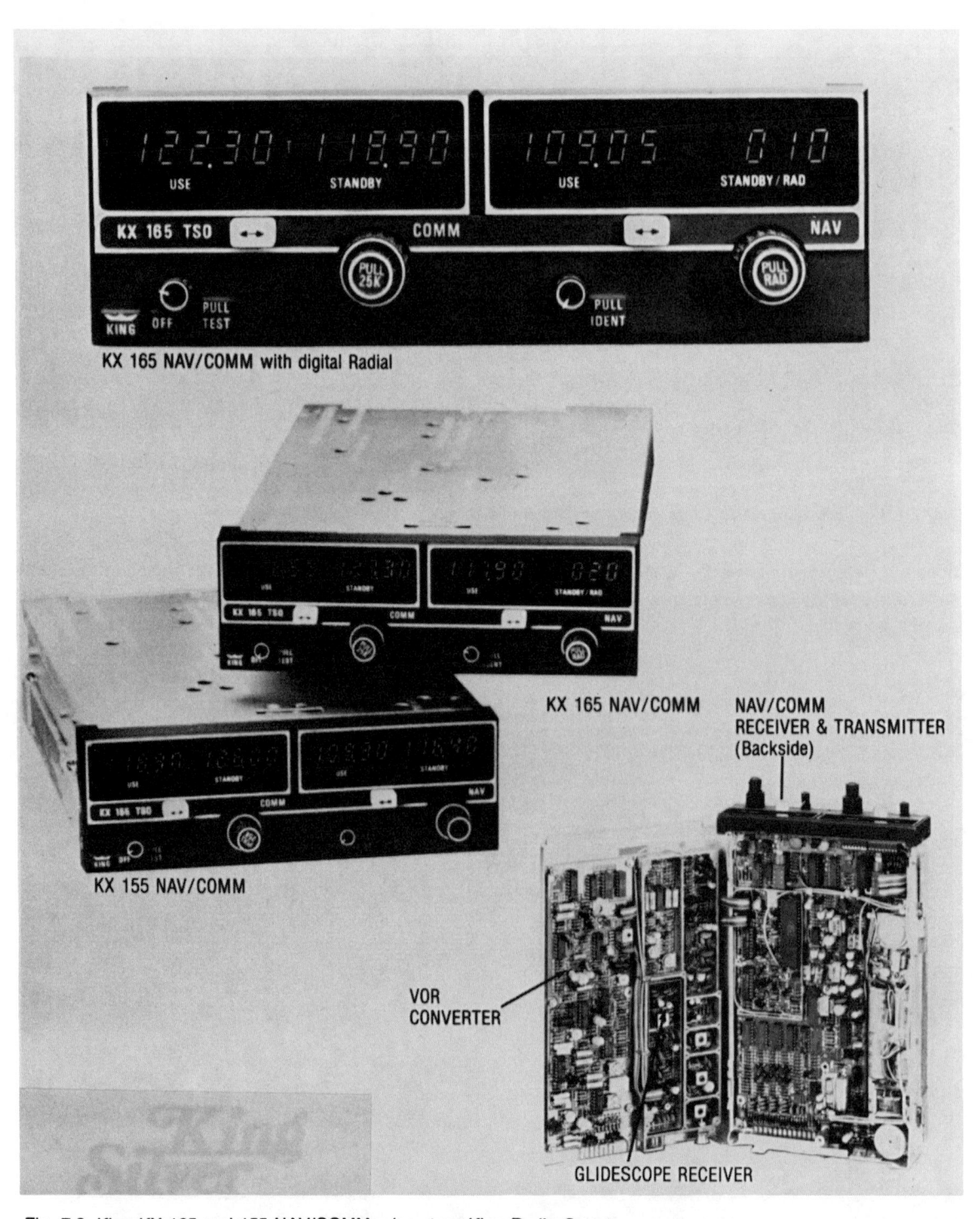

Fig. 7-3. King KX 165 and 155 NAV/COMMs. (courtesy King Radio Corp.)

Fig. 7-4. Narco MK 12D NAV/COMM. (courtesy NARCO Avionics)

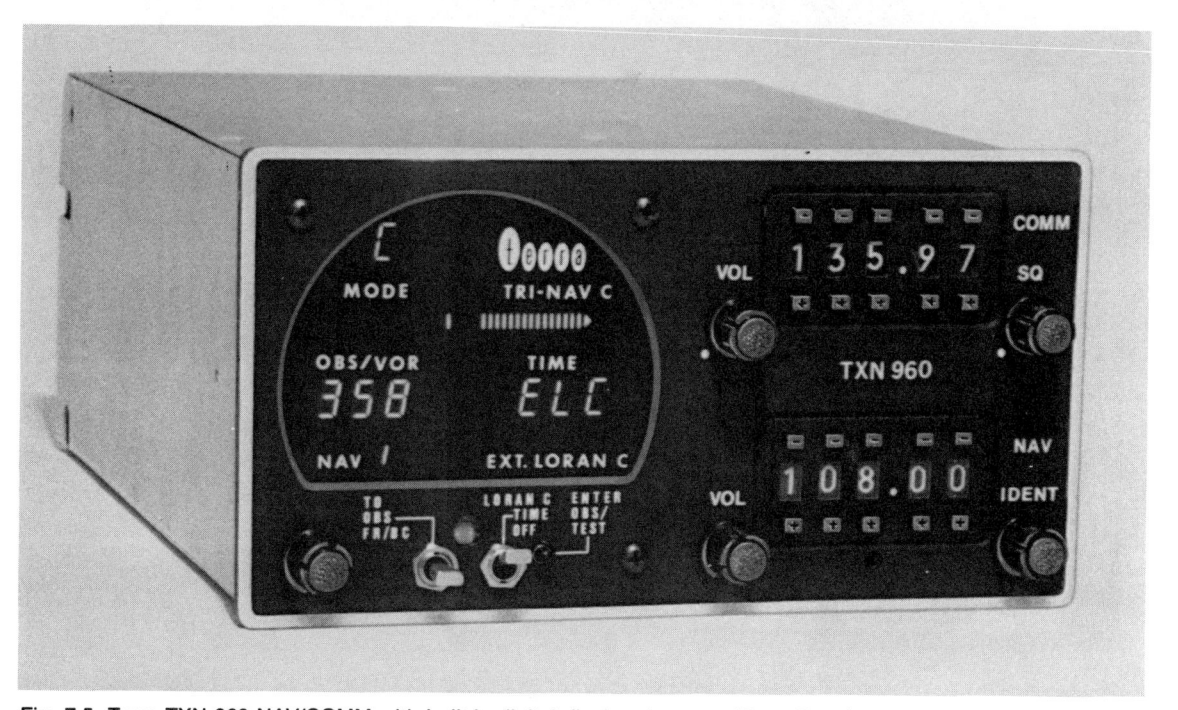

Fig. 7-5. Terra TXN 960 NAV/COMM with built-in digital display. (courtesy Terra Corp.)

55

Fig. 7-6. Narco Escort II NAV/COMM which mounts in a standard instrument panel opening. (courtesy NARCO Avionics)

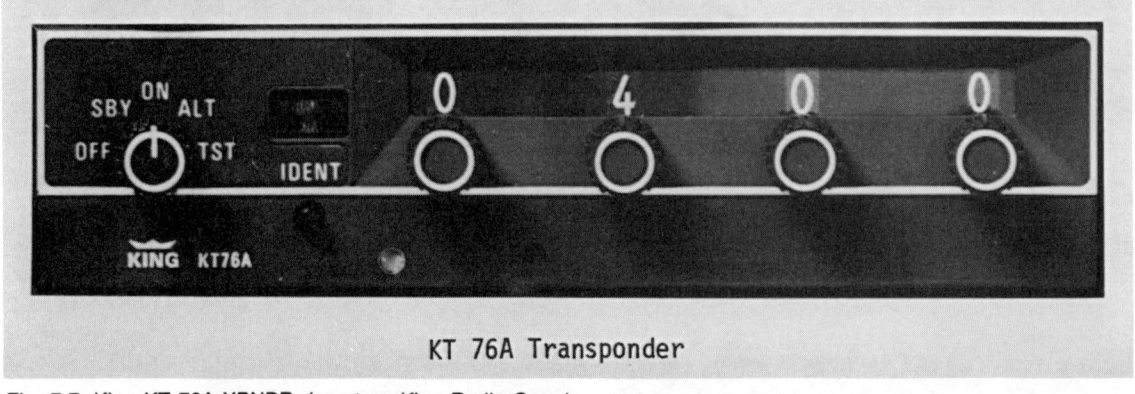

Fig. 7-7. King KT 76A XPNDR. (courtesy King Radio Corp.)

56

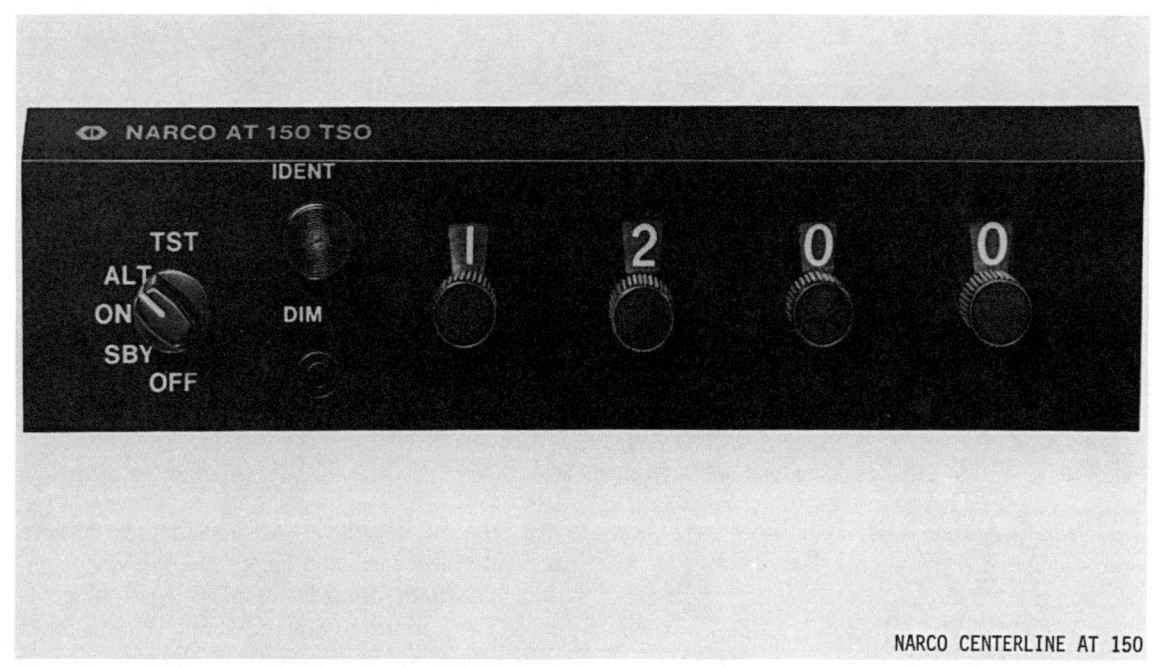

Fig. 7-8. Narco AT 150 XPNDR. (courtesy NARCO Avionics)

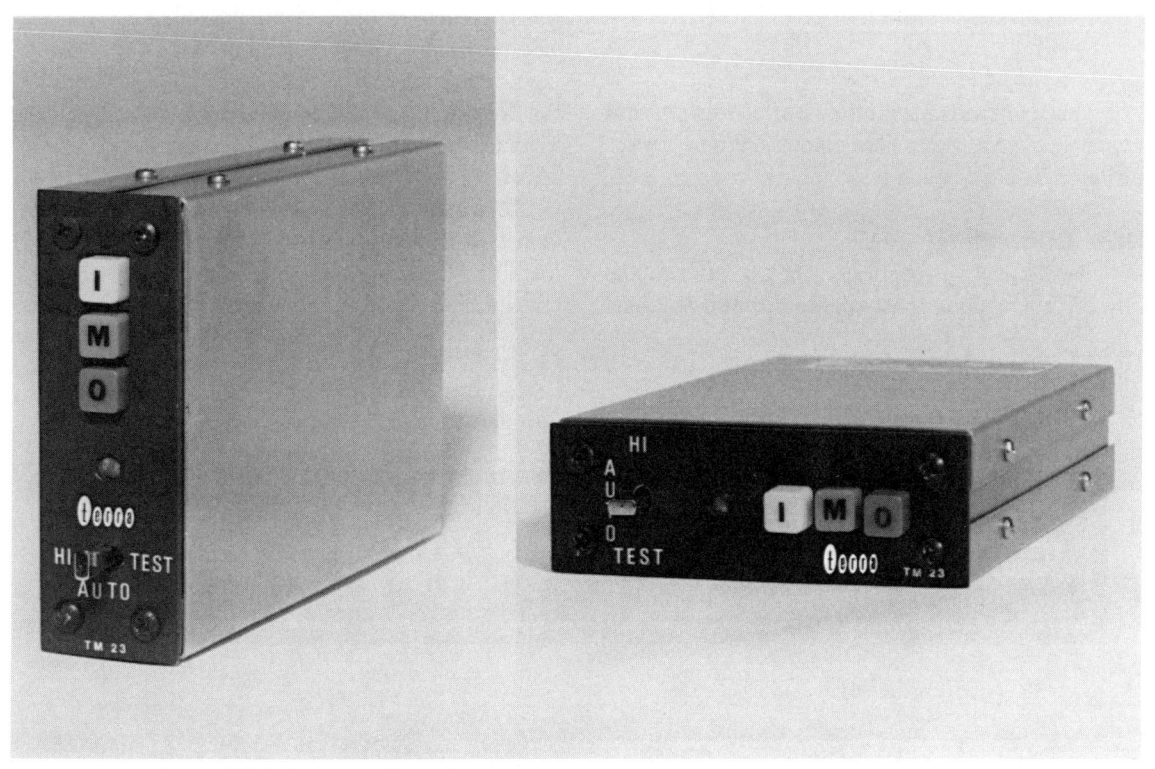

Fig. 7-9. Terra TM 23 MBR. (courtesy Terra Corp.)

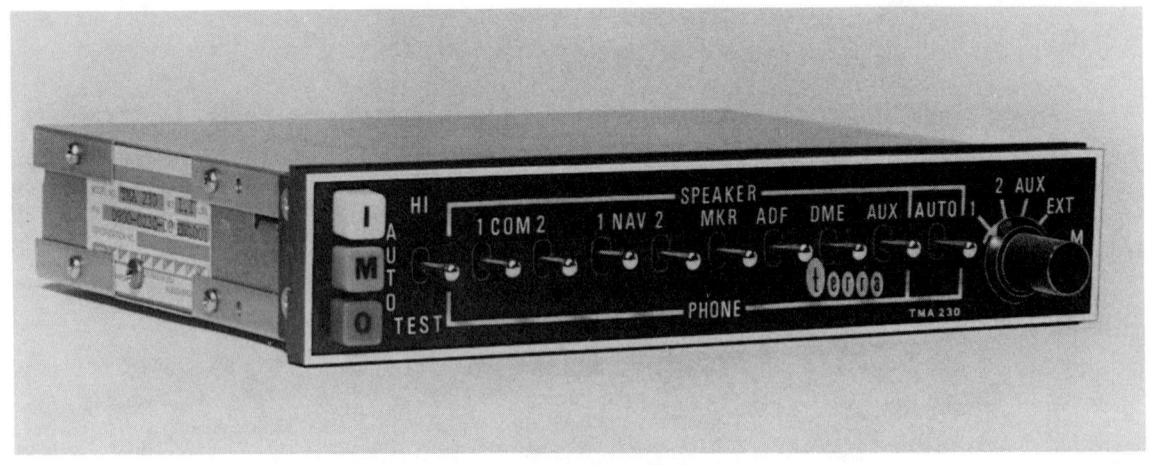

Fig. 7-10. Terra TMA 230 audio panel with MBR. (courtesy Terra Corp.)

If you fly occasional IFR you'll need still more equipment.

- ☐ Dual NAV/COMM
- ☐ LOC/GS/MBR
- ☐ ADF
- ☐ XPNDR w/altitude reporting
- ☐ ELT

There are various methods of acquiring your avionics. You may purchase equipment new, reconditioned, used, or even build your own from a kit.

NEW EQUIPMENT

New equipment is state-of-the-art, offering the newest innovations, best reliability, and smallest size. Some of the units are so small that two or more separate pieces are combined into one very complete package. Considering that panel space is usually at a premium in a homebuilt, new becomes quite attractive. In addition to the newness, there is the warranty factor. Nothing beats a "new" warranty. If these arguments are not enough, then consider this: The latest equipment has the lowest electrical power requirements.

New avionics can be purchased from your local avionics dealer, or thru a discount house (many advertise in *Trade-A-Plane*).

You can visit your local avionics dealer and purchase all the equipment you want, and have him install it. Of course, this will be the most expensive method. However, you'll have new equipment,

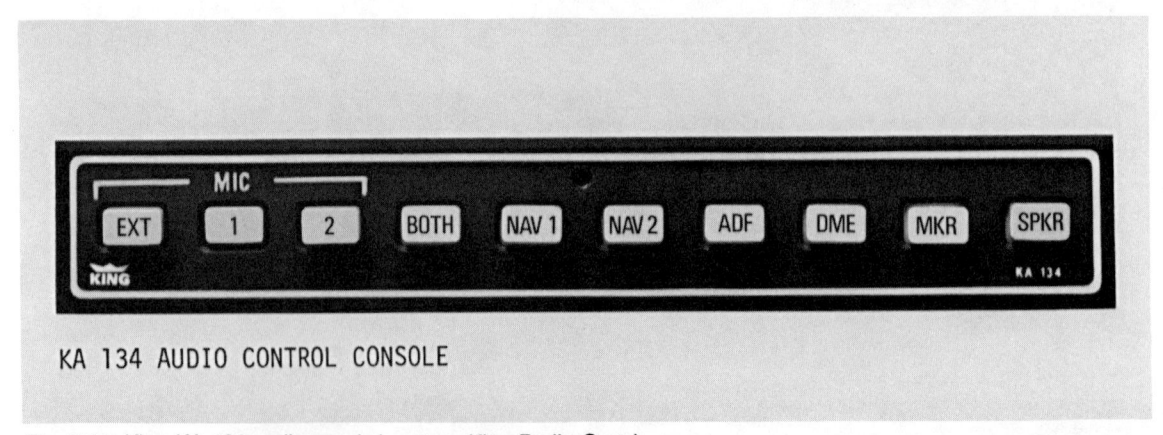

Fig. 7-11. King KA 134 audio panel. (courtesy King Radio Corp.)

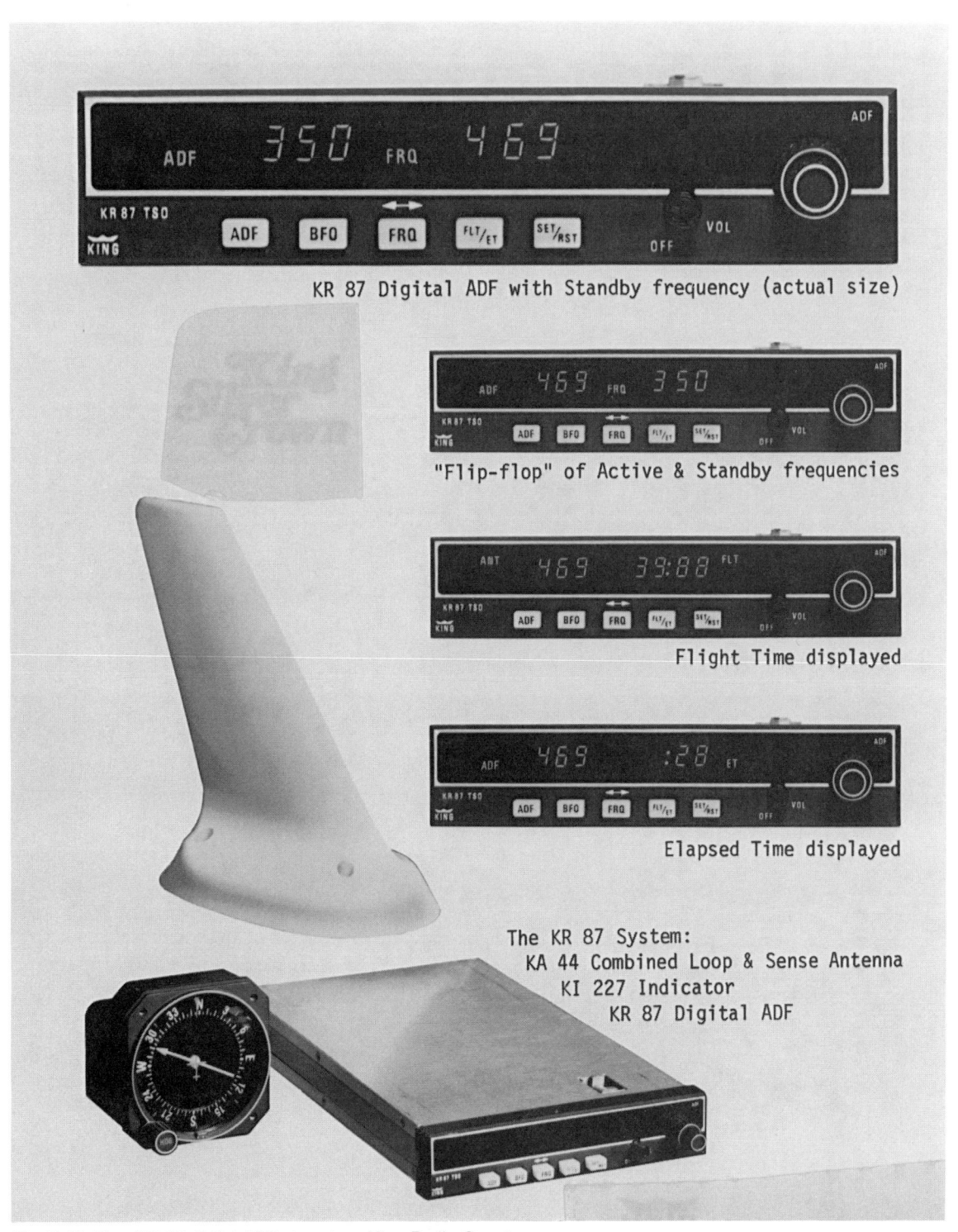

KR 87 Digital ADF with Standby frequency (actual size)

"Flip-flop" of Active & Standby frequencies

Flight Time displayed

Elapsed Time displayed

The KR 87 System:
KA 44 Combined Loop & Sense Antenna
KI 227 Indicator
KR 87 Digital ADF

Fig. 7-12. King KR 87 digital ADF. (courtesy King Radio Corp.)

59

KN 62A DME   Distance/Groundspeed/TTS (GS/T mode)

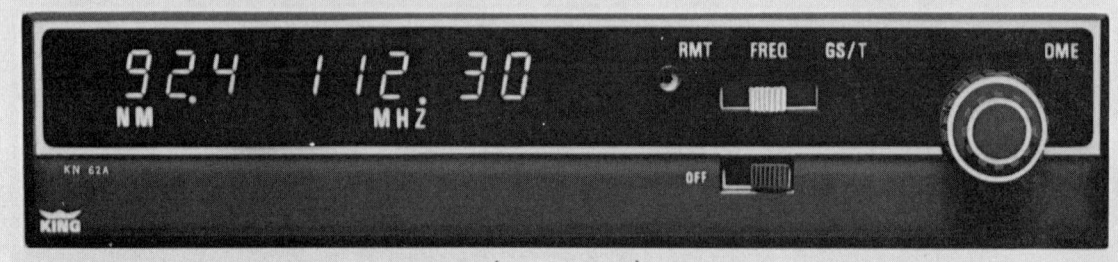

KN 62A DME   Distance/Frequency (FREQ mode)

Fig. 7-13. King KN 62A DME. (courtesy King Radio Corp.)

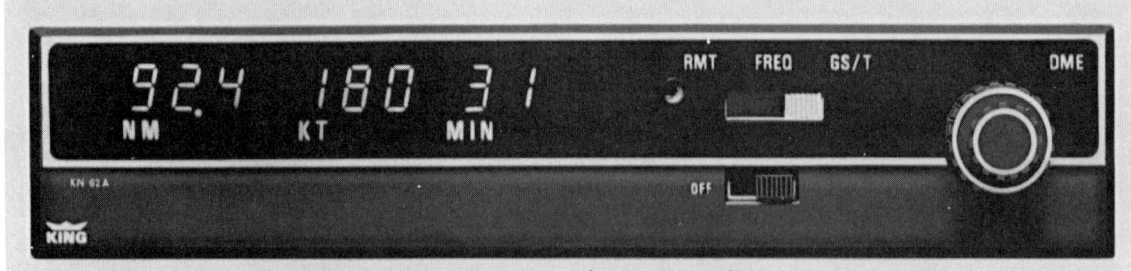

Fig. 7-14. Narco 890 DME. (courtesy NARCO Avionics)

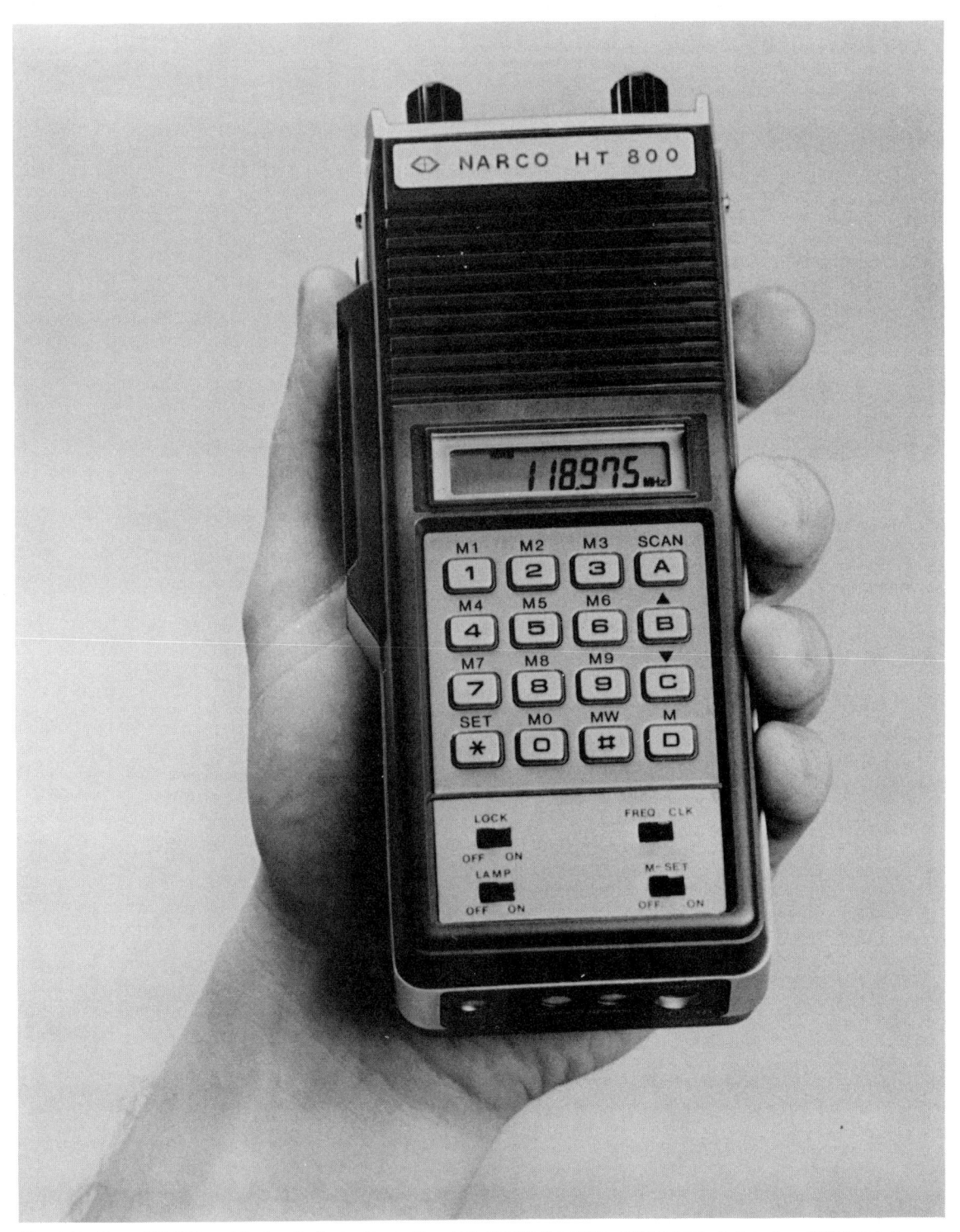

Fig. 7-15. Narco HT 800, an ideal radio for a plane without an electrical system, or as a backup. (courtesy NARCO Avionics)

Fig. 7-16. Communications Specialists TR 720 shown with some of the available accessories. (courtesy Communications Specialists)

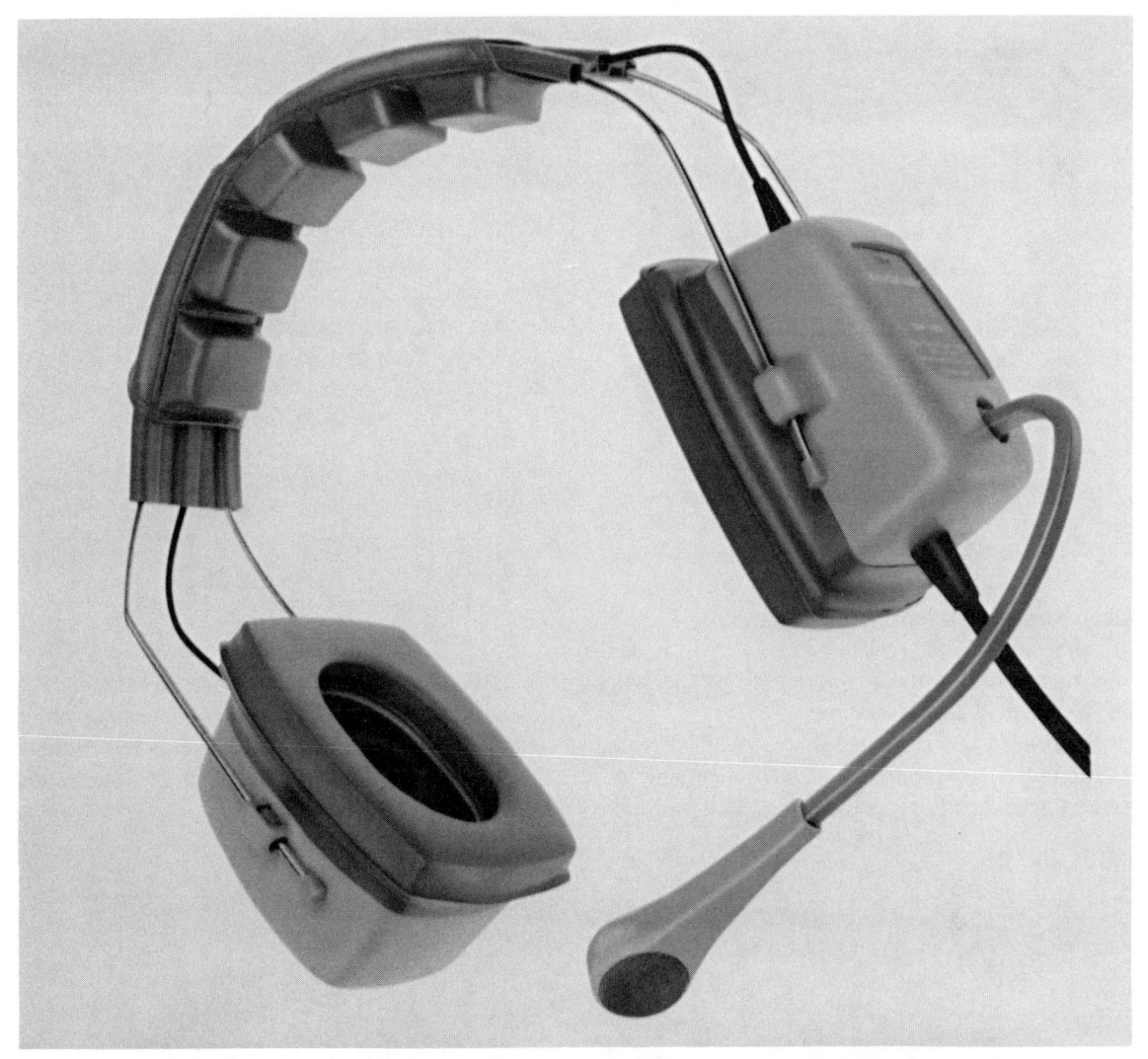

Fig. 7-17. Telex headsets can really reduce noise fatigue and increase ease of radio operation in a noisy airplane. (courtesy Telex)

expert installation, and service backup. I personally wouldn't allow anyone else to touch my airplane but me; however, that's just the way I feel.

The discount house will save you money at the time of the initial purchase, but you may be left out when the need for warranty service arises. Some manufacturers won't honor warranty service requests unless the equipment was purchased from *and installed by* an authorized dealer. This may not seem fair to the consumer, but it is an effective method of protecting the authorized dealers. There may be some flexibility here, as many avionics dealers don't want to "fool around" with homebuilt installations. Be sure to check this out before making a purchase.

RECONDITIONED EQUIPMENT

There are several companies that advertise reconditioned avionics at bargain—or at least low—prices. This equipment has been removed from service and completely checked out by an avionics shop. Parts that have failed, are near failure, or are

63

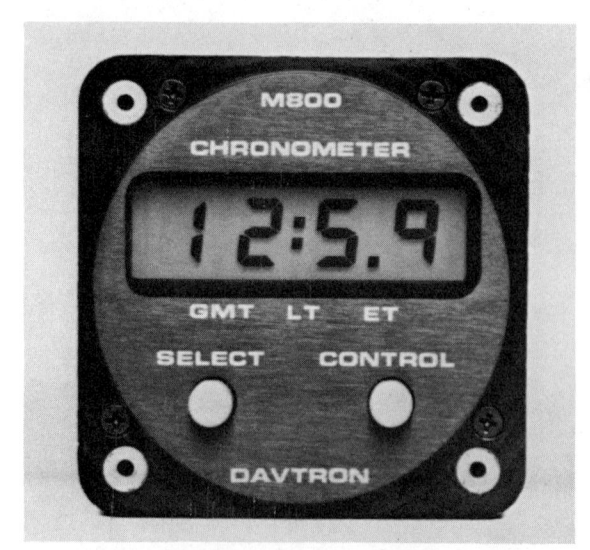

Fig. 7-18. Digital chronograph. (courtesy DAVTRON)

likely to fail, will have been replaced.

These radios offer a fair buy for the airplane owner, and are normally warrantied by the seller for a specified period of time.

However, reconditioned is not *new*. Everything in the unit has been used, but not everything will be replaced during reconditioning. You will have some new parts and some old parts.

If you stick with recent (no more than five or six years old) equipment you should be able to take advantage of some of the modern conveniences found on the latest units.

USED EQUIPMENT

Used avionics can be purchased from dealers or individuals. The aviation magazines and *Trade-A-Plane* are good sources of used equipment.

A few words of advice for those contemplating the purchase of used avionics:

☐ Purchase nothing with tubes in it.
☐ Purchase nothing more than six years old.
☐ Purchase nothing made by a defunct manufacturer.
☐ Purchase nothing "as is."
☐ Purchase nothing "working when removed."

Used equipment can be a wise investment, but it is very risky unless you happen to be an avionics technician, or have access to one. I recommend against the purchase of used avionics, unless you are *very* familiar with the source.

Fig. 7-19. Digital temperature gauge. (courtesy DAVTRON)

RADIO KITS

Now enter the latest method of purchasing avionics—buy kits and build them yourself.

Radio Systems Technology offers a limited line of avionics from audio panels to a 720 channel NAV/COMM in kit form.

For certification purposes, you build it, then ship it back to the manufacturer for checkout and certification. By building it you will save money and learn about the inside of these complex boxes (so says the kit manufacturers). RST will also service it at a later date, should the need arise. This is certainly an interesting way for the budget-minded individual to acquire avionics (See Fig. 7-20).

For further information about RST kits contact:

Radio Systems Technology
13281 Grass Valley Ave.
Grass Valley, CA 95945
Phone: (916) 272-2203

ANTENNAS

On the normal factory-built all-metal airplane you have no doubt noticed all the antennas that are mounted on the fuselage and wings. On a composite you can place these airflow-disturbing devices internally. After all, plastics don't conduct, and radio waves pass through them with no loss or signal degradation.

This brings about an interesting point: Composite airplanes are not good radar targets. The materials they are built of do not reflect radio waves very well. This is just another reason for having a transponder. With it you'll make a larger blip on some FAA controller's screen, and might be a little safer.

Antenna Dynamics recently introduced a NAV/LOC/GS antenna designed for "within structure" installation. The antenna comes complete, ready for installation, and with very good instructions (reprinted here by permission of Antenna Dynamics):

Installation Instructions
for
Navigation/localizer/glidescope antenna
Part Number AD-1

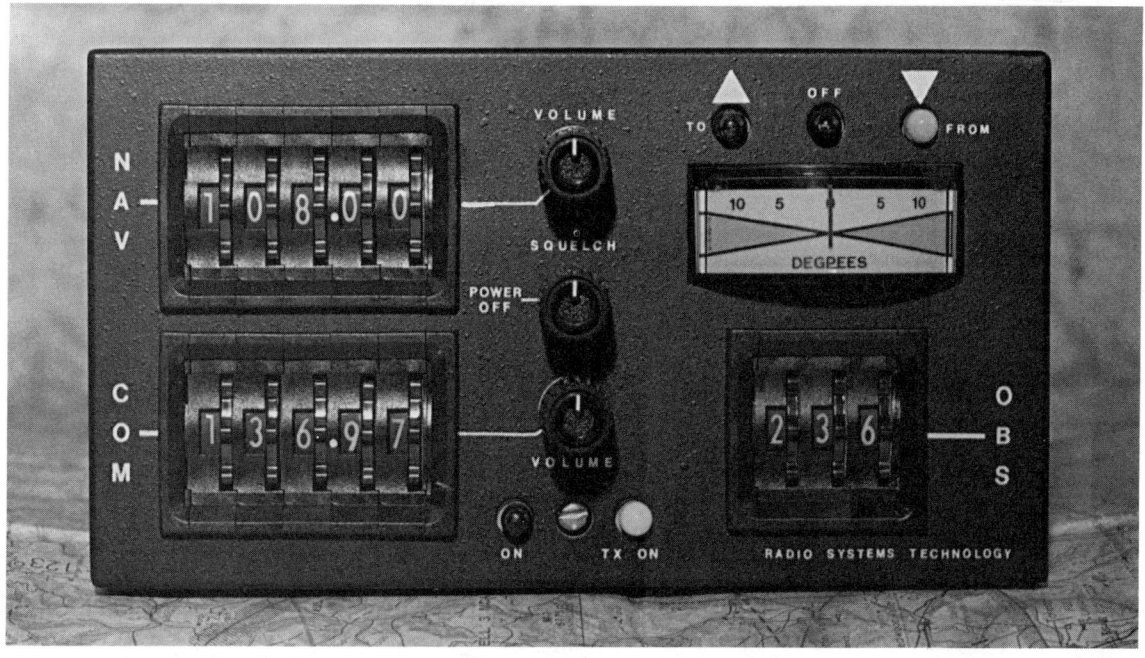

Fig. 7-20. RST's NAV/COMMs can be assembled in about 60 to 80 hours, and are available factory-direct in either 360 channel (RST-571) or 720 channel (RST-572) versions. (courtesy RST)

1. This sheet describes in general terms the installation of the so-called "hidden" (also referred to as "flush" or "conformal") antenna systems on foam-fiberglass "plastic" airplanes.

2. The p/n AD-1 is designed to "look out" (receive) through foam-fiberglass (non-metallic) surfaces. It cannot "see" (receive) the incoming NAV/LOC/GS signals when mounted directly on heavy wood, metal, or carbon cloth materials. Care should be taken to choose a large enough horizontal area so the antenna is not guarded (shielded) by metallic surfaces either directly under the antenna or within the immediate area of installation.

3. The p/n AD-1 NAV/LOC/GS antenna is basically a copper foil element dipole fed through a unique state-of-the-art micro electronic matching device that allows maximum reception sensitivity (commonly referred to as gain) throughout both the Navigational/Localizer frequency range of 108-118 MHz and the Glideslope frequency range of 329-335 MHz. The basic antenna design is now being used (in different configurations) on many type aircraft up to and including business jets, assuring its high degree of capability and reliability.

4. The antenna may be installed in standard dipole configuration (Fig. 7-21) or in "V" configuration. Remove backing tape from copper elements and install per Fig. 7-21. Be careful not to bend lead wires too many times as they may work-harden and break.

5. The antenna may be placed anywhere in or on the foam or fiberglass; the surface, inside the glass but on the foam, or buried within the foam are all valid locations for the antenna. In order to optimize antenna performance, try to keep the antenna as close as possible to the bottom surfaces of the aircraft (Fig. 7-22).

6. Normally all NAV/LOC/GS antennas are installed horizontal to the earth's surface as this is the electrical plane in which the NAV/LOC/GS signals are both sent from the stations and received by your antenna. A little thought will show that your antenna can be placed in a large number of locations in most plastic airplanes (i.e., canard, main wing, fuselage, etc.). Choose the most suitable configuration for your antenna. If the whole antenna will not fit flat inside the horizontal surface, the foil elements may be slightly bent to conform to the airframe surface.

7. Install the antenna as far as possible away from metal surfaces or other wires or electrical devices.

8. Remember, if the airplane were made out of clear glass, and if a person at the NAV/LOC/GS station couldn't "see" the antenna (and that includes looking through engines, fuel tanks, people, etc.), then the antenna will not be able to properly receive the Navigational/LOC/GS signals (Fig. 7-23).

9. Do not install the antenna in areas of high flex. An analysis of the circumstances surrounding failures in high-flex areas leads to the conclusion that antennas installed on fiberglass surfaces subject to flex are most likely to break. Although the tape is really quite strong, it cannot survive the strain imposed by a half-ton airplane bouncing along the runway. All of the reported failures have been on gear leg antennas or canard gear antennas, especially after hard landings. Remember that copper foil tape is not structural, and that the foil tape is much more likely to fail on tension than any other mode.

10. The antenna may be used for VOR, VOR/LOC, Glideslope or any combination of these functions. However, if multiple functions are to be used a "splitter" box (readily available from your avionics shop) must be used (Fig. 7-24).

11. It is virtually impossible for us to answer all the questions you may have as this antenna system is meant for use on many types of aircraft. When in doubt, consult with the avionics man at your local airport or question other builders of aircraft.

12. Statistics on thousands of hours of antenna flight show that approximately 90 percent of all antenna problems are due to improper installation.

13. Keep the antenna as far away as possible from other antennas (especially VHF communications antennas).

**STANDARD CONFIGURATION
FRONT OF FUSELAGE**

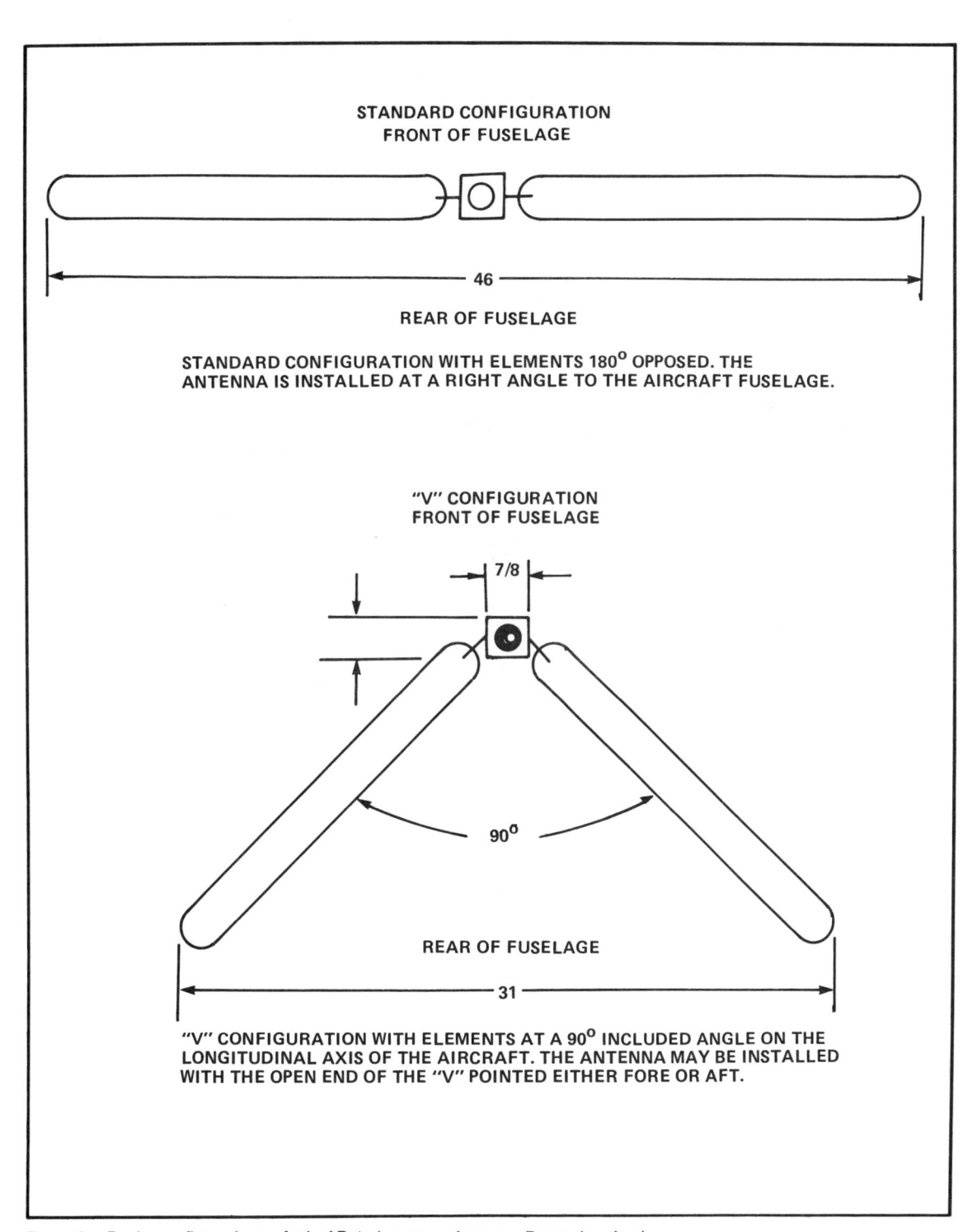

|←————————— 46 —————————→|

REAR OF FUSELAGE

STANDARD CONFIGURATION WITH ELEMENTS 180° OPPOSED. THE ANTENNA IS INSTALLED AT A RIGHT ANGLE TO THE AIRCRAFT FUSELAGE.

**"V" CONFIGURATION
FRONT OF FUSELAGE**

|← 7/8 →|

90°

REAR OF FUSELAGE

|←————————— 31 —————————→|

"V" CONFIGURATION WITH ELEMENTS AT A 90° INCLUDED ANGLE ON THE LONGITUDINAL AXIS OF THE AIRCRAFT. THE ANTENNA MAY BE INSTALLED WITH THE OPEN END OF THE "V" POINTED EITHER FORE OR AFT.

Fig. 7-21. Basic configurations of p/n AD-1. (courtesy Antenna Dynamics, Inc.)

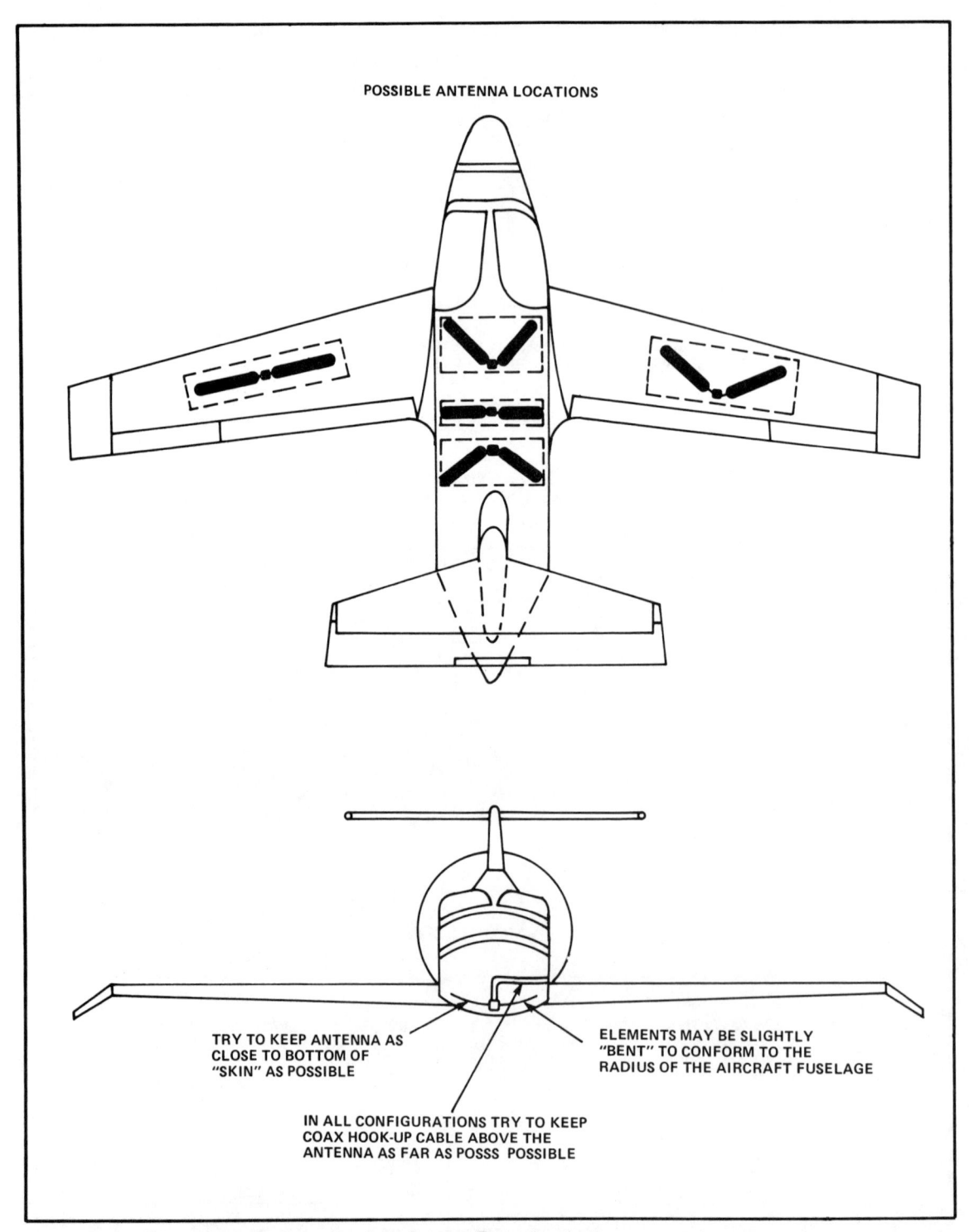

Fig. 7-22. Possible antenna locations. (courtesy Antenna Dynamics, Inc.)

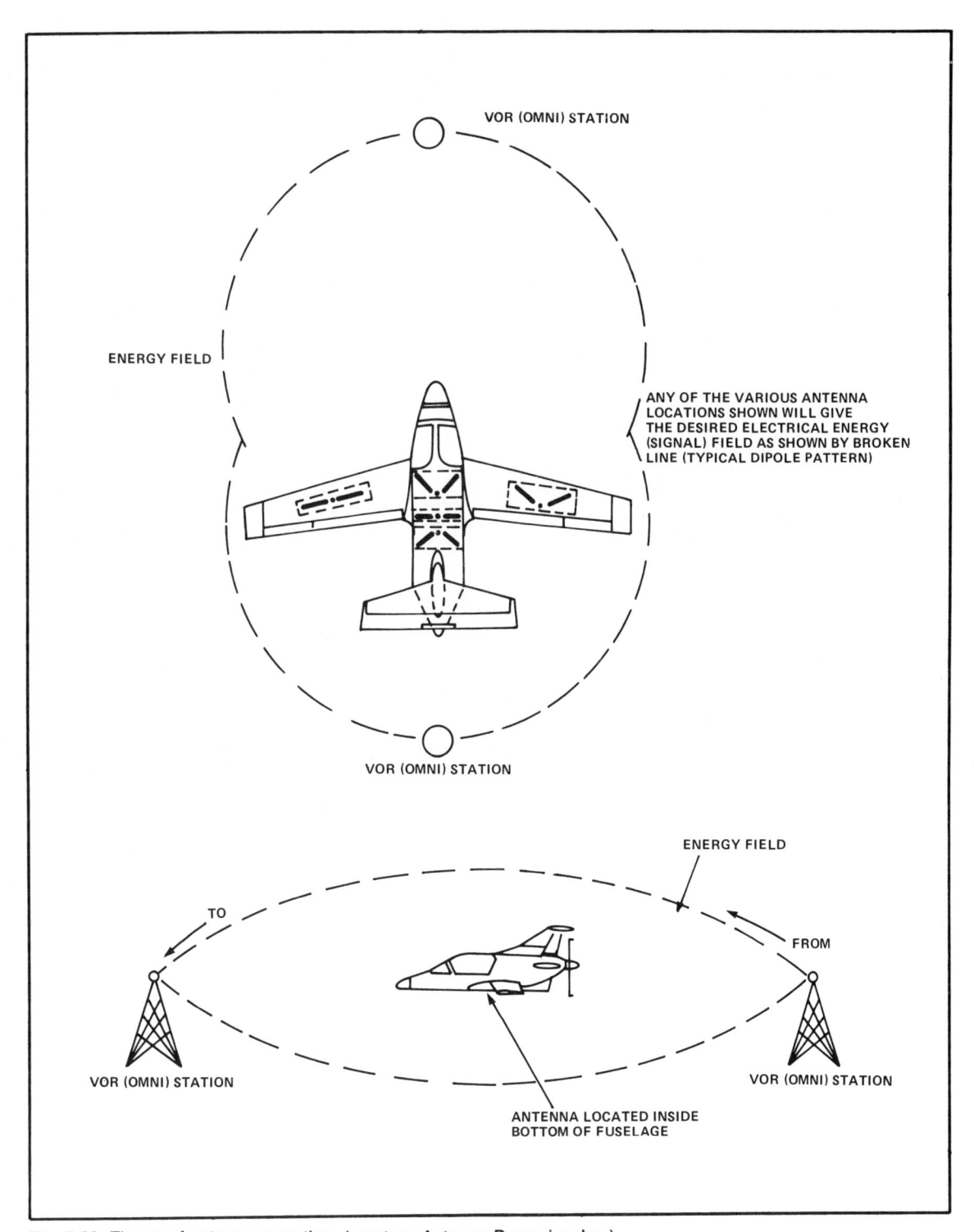

Fig. 7-23. Theory of antenna operation. (courtesy Antenna Dynamics, Inc.)

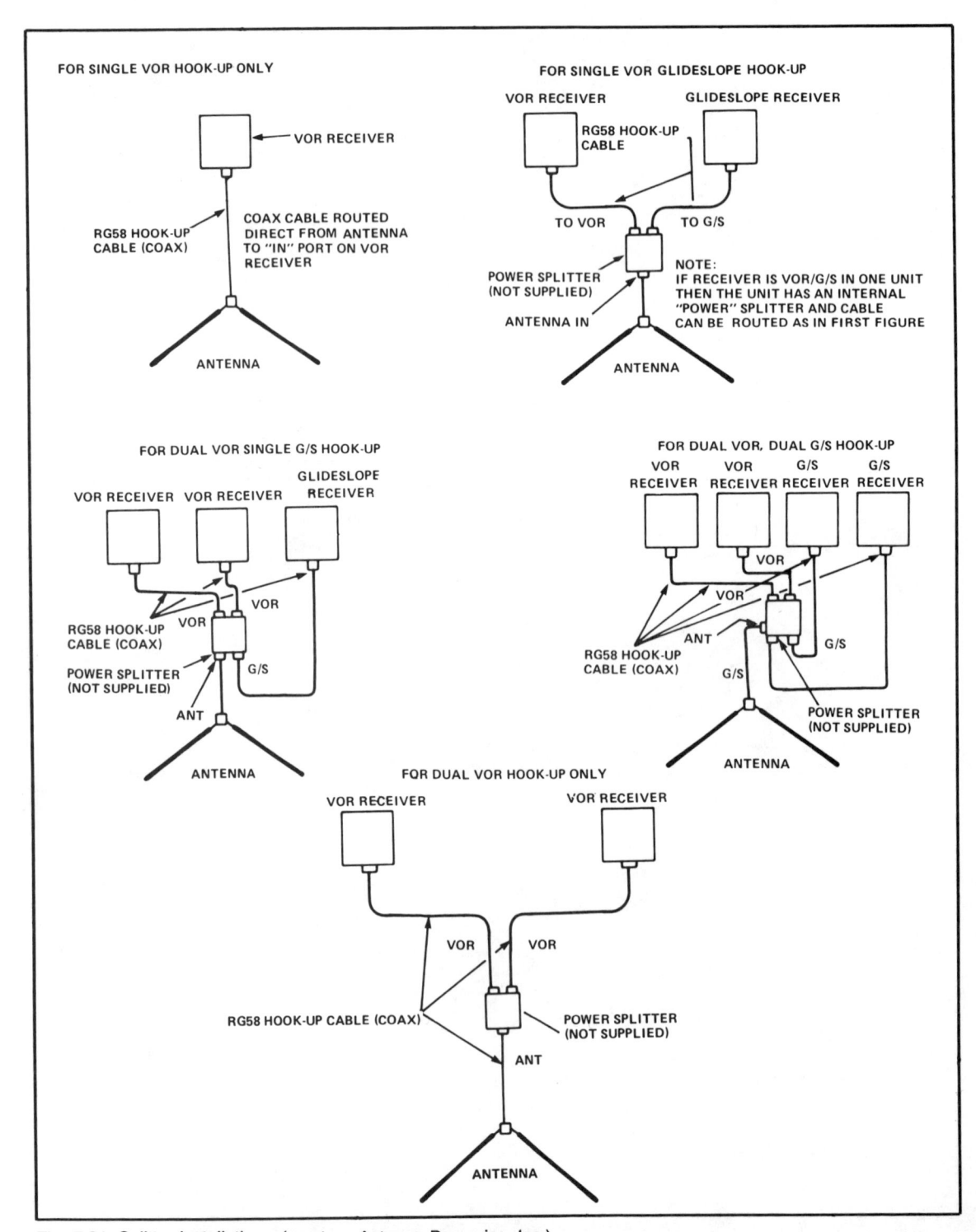

Fig. 7-24. Splitter installations. (courtesy Antenna Dynamics, Inc.)

SELECTING AN INSTRUMENT GROUPING

1. ESTABLISH REQUIREMENTS—BASIC PANEL WITH NO GYRO INSTRUMENTS, WITH GYROS, OR FULL IFR.
2. NATURAL GROUPING—FLIGHT INSTRUMENTS (AIRSPEED, ALTIMETER, RATE-OF-CLIMB), ENGINE INSTRUMENTS (TACH, OIL PRESSURE, CYL. HEAD TEMP., ETC.)
3. NORMAL SCAN—PROVIDE LAYOUT CONDUCIVE TO NORMAL LEFT TO RIGHT SCAN.
4. KEEP COMPASS OUT OF PANEL—AVOID PLACING INSTRUMENTS WITH ELECTRICAL METER MOVEMENTS NEAR COMPASS.
5. BE SURE YOU HAVE SUFFICIENT DEPTH BEHIND PANEL BEFORE YOU CUT THE PANEL. SOME INSTRUMENTS SUCH AS GYROS AND RATE-OF-CLIMBS CAN BE OVER 8 INCHES LONG, NOT INCLUDING SPACE REQUIRED BY THE LINES AND FITTINGS.
6. WHEN PLUMBING GYROS, USE LARGE RADIUS TURNS.
7. AVOID ANY CONTACT BETWEEN GYRO INSTRUMENTS.

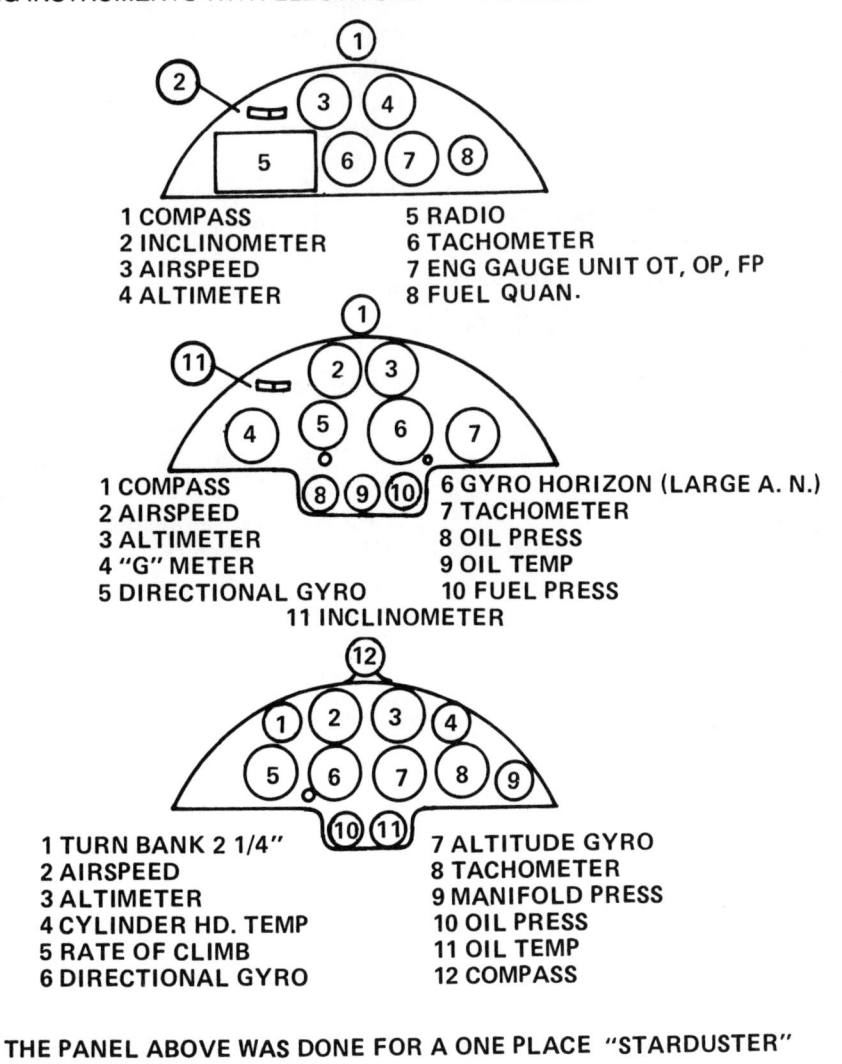

1 COMPASS
2 INCLINOMETER
3 AIRSPEED
4 ALTIMETER
5 RADIO
6 TACHOMETER
7 ENG GAUGE UNIT OT, OP, FP
8 FUEL QUAN.

1 COMPASS
2 AIRSPEED
3 ALTIMETER
4 "G" METER
5 DIRECTIONAL GYRO
6 GYRO HORIZON (LARGE A. N.)
7 TACHOMETER
8 OIL PRESS
9 OIL TEMP
10 FUEL PRESS
11 INCLINOMETER

1 TURN BANK 2 1/4"
2 AIRSPEED
3 ALTIMETER
4 CYLINDER HD. TEMP
5 RATE OF CLIMB
6 DIRECTIONAL GYRO
7 ALTITUDE GYRO
8 TACHOMETER
9 MANIFOLD PRESS
10 OIL PRESS
11 OIL TEMP
12 COMPASS

THE PANEL ABOVE WAS DONE FOR A ONE PLACE "STARDUSTER"

Fig. 7-25. Selecting an instrument grouping. (courtesy Aircraft Spruce and Specialty Co.)

For further information about these antennas contact:

Antenna Dynamics, Inc.
1251 W. Epulveda Blvd.
Suite #268
Torrance, CA 90502
Phone: (213) 534-1090 ext. 22

You will find additional instructions for mounting transmitting and receiving antennas in the plans for most kits that are described in this book. All of these antennas can be embedded within the structure.

INSTRUMENTATION

Now that the question of radios has been discussed, there is still the remainder of the instrument panel to be taken care of. There is quite a bit of freedom here, as long as the FAA is kept happy with the installation of certain required instruments. These are:

- ☐ Magnetic compass.
- ☐ Tachometer.
- ☐ Oil pressure gauge.
- ☐ Oil temperature gauge.
- ☐ Fuel quantity.
- ☐ Airspeed indicator.
- ☐ Altimeter.

In addition to these, if you wish to fatten up the panel for IFR you will add:

- ☐ Directional gyro.
- ☐ Sensitive altimeter.
- ☐ Attitude indicator.
- ☐ Turn and bank indicator.
- ☐ Clock.

As you can see from this list, much time and money are going to be spent on that instrument panel (Fig. 7-25).

I have seen a few exceptional panels recently completely set up for IFR, with everything duplicated for the copilot—except that the pilot's instruments were electric, and the copilot's were vacuum. This combination system is perfect for complete redundancy. It is also very costly, to say nothing of the time and effort to build and install. At the other end of the scale I have seen bare minimum panels, with a hand-held COMM radio laying on the seat. Who is to say who is having more fun?

Each homebuilt airplane is a dream of the builder; where his dreams carry him will determine how complex or simple the instrumentation and communications will be on his plane.

SUMMARY

The composite airplanes are state-of-the-art, smooth, sleek, and fast. I feel that to do them justice you should have state-of-the-art equipment on board. With this in mind, and many years of experience in electronics, I strongly recommend that when contemplating the purchase of avionics and/or instruments, *save your money until you can purchase new equipment.*

Recommended reading: *Upgrading Your Airplane's Avionics*, TAB Book No. 2301, by Timothy R.V. Foster.

Chapter 8

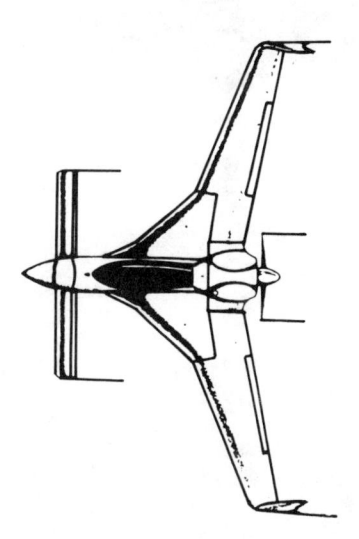

Popular Composite Airplanes

The following pages contain descriptions, specifications, and photographs of the various composite airplanes available on today's market.

AERO MIRAGE

The Aero Mirage TC-2 is a two seat (side-by-side) retractable tri-gear airplane built from kit (premolded) parts (Fig. 8-1). Extensive use of Kevlar is made in this design for both strength and lightness.

The TC-2 is designed to use the Continental 0-200 engine. However, any engine from 65 hp to 100 hp can be used (with slightly reduced performance figures). The 0-200 engine is readily available and is a proven, reliable powerplant. Also, the O-200 is a fuel miser, yet when installed on the TC-2 can provide speeds in excess of 200 mph.

Specifications for the Aero Mirage TC-2 (100 hp):

Dimensions
  Length: 16.6 ft.
  Span: 21.0 ft.
  Wing area: 64 sq. ft.

Weights
  Empty: 640 lbs.
  Gross: 1140 lbs.
  Useful load: 425-500 lbs.
  Baggage: 40 lbs.
  Fuel cap: 25 gal.
  Fuel usage: 36-42 mpg.

Performance
  Cruise   (75%):   190 mph.
           (100%):   208 mph.
  V_{ne} 259 mph.
  Stall   (w/flaps):   57 mph.
           (clean):   63 mph.

Construction

Although the Aero Mirage TC-2 can be built from the kit at home, the factory recommends you spend a few days with them and assemble some of your aircraft under their supervision. This is especially recommended for first-time builders. Us-

Fig. 8-1. Aero Mirage. (courtesy Aero Mirage Aircraft)

ing this method, most builders can get their entire structural airframe completed in only one week. The factory is in "sunny" Florida.

For further information contact:

Aero Mirage, Inc.
3009 N.E. 20th Way
Gainsville, FL 32609
Phone: (904) 377-4146

COZY

The Cozy (Fig. 8-2) was developed by Nat and Shirey Puffer, originally as a one-of-a-kind modification to their Long-EZ. It was first flown in July of 1982.

As described in the advertising literature: The Cozy is a small, compact, high-performance, high-utility sportplane featuring side-by-side seating for an average-size couple or smaller, full dual controls, an expansive instrument panel, and a large baggage bay in the rear which could double as an extra seat. While recommended mainly for day VFR operations, competent pilots can also equip it for night and IFR flying.

The recommended powerplant is any model of the O-235 Lycoming. A mechanical fuel pump is required. It has an alternator-powered electrical system and can be equipped with an electric starter.

The cockpit layout is designed to complement pilot and/or copilot work load, with throttle, mixture, carb heat, pitch trim, aileron trim, landing brake, landing light, nosewheel crank, cabin heat, and fuel tank selector valve located in a center console for equal access to both pilot and copilot, and individual side-stick controllers on both outside armrests. Seating provides armrest, lumbar, thigh, and head support for "recliner-chair" comfort not found in conventional aircraft seats. This allows long, fatigue-free flights.

The Cozy uses the very latest aerodynamic technology, combining winglets, a high aspect-ratio wing with Eppler airfoils optimized for efficient cruise, and a configuration with far less wetted area than conventional airplanes. Because its power-off glide angle is only 3.7 degrees, a belly-mounted landing brake is used to steepen the descent to landing.

The flying qualities of the Cozy are superb. It

is a very solid, stable airplane that has responsive ailerons, good turbulence resistance, excellent "hands-off" stability, and docile stall characteristics. It resists stalls and spins, even when maneuvered sharply to full aft stick. Flight tests show the prototype to be free from stall departures and spins. Climb is excellent, even at full aft stick speed. Trim changes due to power, gear retraction, or landing brake are very small.

As with all pusher type airplanes of this type, operation from unpaved runways is not recommended.

Specifications for the Cozy):

Dimensions
- Span: 26.1 ft.
- Wing area: 95.6 sq. ft.

Weights
- Empty: 850 lbs.
- Solo: 1100 lbs.
- Gross: 1500 lbs.

Fuel cap: 52 gal.

Cabin
- Length: 100 in.
- Width: 40 in.
- Height: 38 in.

Performance
- Takeoff (solo): 670 ft.
- (gross): 1050 ft.
- Climb (solo): 1500 fpm.
- (gross): 900 fpm.
- Cruise (75%): 180 mph.
- (40%): 143 mph.
- Fuel usage (75%): 6.7 gph.
- (40%): 3.6 gph.
- Range (75%): 1200 mi.
- (40%): 1800 mi.

Fig. 8-2. Cozy. (courtesy Co-Z Development Corp.)

Duration	(40%):	12.8 hrs.
		(w/1.5 hr. reserve).
	(75%):	7 hrs.
		(w/1 hr. reserve).
Ceiling	(solo):	25,000 ft.
	(gross):	20,000 ft.
Landing	(solo):	550 ft.
	(gross):	950 ft.
Approach speed:		80 mph.
Landing speed:		65 mph.

Construction

The typical building time for the Cozy is 1700 hours; however, this can be reduced by the use of prefabricated parts available from various suppliers. The reduced time could be as little as 1200 hours; however, the cost will be somewhat higher.

A monthly newsletter is published by the manufacturer. Its purpose is to keep builders abreast of updates and changes in the Cozy construction/plans.

For further information contact:

Co-Z Development Corp.
2182 North Payne Ave.
St. Paul, Minn 55117
Phone: (612) 776-1145

DRAGONFLY

The Dragonfly is a two-place canard craft, built in true composite style. It is designed to use the H.A.P.I. model 60-2DM Volkswagen conversion engine.

The original Dragonfly (Fig. 8-3) is one of the most popular homebuilt aircraft designs in the world. Hundreds are currently being built, and there are well over 100 examples already flying.

The original version is best suited to use on wide paved runways, and taxied on wide taxiways. This is due to the very wide stance of the main landing gear, which is actually attached at the ends of the canard.

To fit the Dragonfly to the needs of more pilots, the Mark II was recently introduced. The Mark II has redesigned landing gear with an eight-foot tread, increased canard area for reduced wing loading, and increased elevator area (Fig. 8-4). It has all the speed and efficiency of the original version plus a lower landing speed, 50 mph, and a shorter landing roll, 750 feet. It is quite at home on grass or dirt strips, and operates with the same VW conversion engine as the original version.

To further interest even more pilot/builders, the Mark III was introduced (Fig. 8-5). It is a tri-geared version of the Mark II. On all versions, the major parts are the same, and all can be built from scratch or prefab kits.

The Dragonfly received the "Outstanding New Design Award" at Oshkosh in 1980.

Specifications for the Dragonfly/Mark II/Mark III:

Dimensions
Length:	20.0 ft.
Wing span:	22.0 ft.
Wing area:	97.0 sq. ft.

Weights
Empty:	605 lbs.
Gross:	1075 lbs.

Performance
Takeoff run:	450 ft.
Stall:	45 mph.
Cruise speed:	165 mph.
Range:	500 mi.
Climb (gross):	850 fpm.
Svc ceiling:	18,500 ft.
Engine:	1834cc VW
Fuel usage:	3.4 gph.

Construction

The Dragonfly can be scratch-built from plans, or from prefabricated parts. Basic construction time is claimed to be 1200 hours if no prefab parts are used. This time can be reduced to as little as 500 hours by using all the available prefabricated parts.

If you wish to build really quickly you can purchase your kit and build your airframe in only two weeks. This work is done at the H.A.P.I. Flight Center, using their tools, guidance, and help. When you complete your work at the center you'll leave with wings, fuselage, and control surfaces, and be

Fig. 8-3. Dragonfly I. (courtesy H.A.P.I. Engines)

Fig. 8-4. Dragonfly II. (courtesy H.A.P.I. Engines)

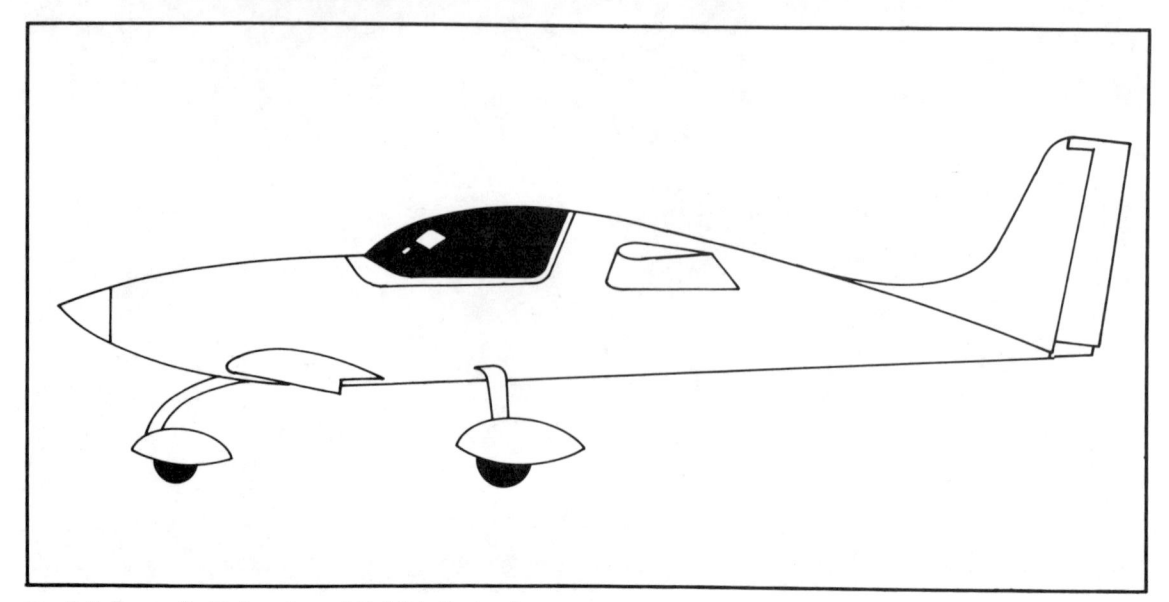

Fig. 8-5. Dragonfly III. (courtesy of H.A.P.I. Engines)

flying in less than 250 hours of additional work. For further information contact:

>H.A.P.I. Engines, Inc.
>Rt. 1 Box 1000
>Eloy, AZ 85231
>Phone: (602) 466-9244

GLASAIR

The Glasair won the EAA Outstanding New Design award at Oshkosh in 1981. Since that time it has become one of the most popular homebuilt airplanes (Fig. 8-6).

As described in the sales literature: From the very beginning the idea behind the Glasair was to create a fast airplane. The design had to be clean, conventional, and create a good-looking airplane that still had a lot of utility—the kind of airplane that everyone who flies dreams of owning. With the costs of operating an aircraft growing daily, economy of operation was also high on the design priority list.

Many high-performance aircraft on the market have trouble at the low end of the speed envelope with high stall speeds and hot approaches needed for safe landings. High stall speeds and hot approaches make landing and takeoff distances longer, which in turn make short, rough fields off-limits for many high-performance planes. Short rough-field capability was a must for the Glasair. The Glasair achieved the best of both worlds, with the design obtaining great performance at both the high and low ends of the speed envelope.

Aerobatic flight was also a consideration of the

Fig. 8-6. Glasair. (courtesy Stoddard-Hamilton Aircraft, Inc.)

Glasair design. This required an airplane with the structural integrity to handle aerobatic flight. The Glasair has been designed to meet these requirements. Structural limit loads for the Glasair under the aerobatic category are +6 Gs and −4 Gs.

To obtain a fast airplane, aerodynamic drag must be conquered. Many things must be considered, such as skin friction drag, pressure drag, surface irregularities, interference drag, lift drag, etc. To attain a low drag profile, critical attention must be paid to the contour and shape of the fuselage. The entire fuselage was computer-lofted including everything from the sleek cowling up front to the perfectly tapered tail at the rear.

From computer lofting to wing design, the Glasair's drag is reduced considerably. To further reduce drag on the taildragger, speed improvements have been made through faired gear legs, wheel pants that also enclose the brakes, fairings that house the wing roots, internal antennas, tailwheel fairings, and control surfaces that leave barely a gap.

For the ultimate in drag reduction, the Glasair RG with the gear retracted is unexcelled.

The Glasair is designed to utilize the very available Avco Lycoming O-320 series engine. All models 150 and 160 hp are approved except the H series. Even the IO-320 may be used. It is available in either taildragger configuration or as a tri-geared retractable.

A Pilot Report

"Three years ago I sold my Bonanza due to rapidly rising fuel and maintenance costs. I wanted an aircraft to do what the Bonanza would do, at a much lower cost. Then I discovered the Glasair. It offered just what I was looking for, so I bought one of their first kits. I gave up two seldomly-used back seats for double the gas mileage, and by being the builder, I could do all the maintenance myself and cut costs drastically.

"The construction looked quick and easy to build. It has unbeatable speed and economy and a high enough wing loading to give a good ride at 200 mph plus, and it still has good short and rough field capabilities to handle the back-country strips of Idaho where I go every year. I also like the side-by-side seating for convenience and especially CG control allowing for good baggage capacity (tent and sleeping bags).

"I have flown over 150 makes and types of aircraft the past 24 years, including all three Glasair prototypes, and I must say I put the Glasair on the front page of my flying favorites. It has the easy slow-flight handling capability of cruising in formation with Champs and Cubs (sliding window open and elbow out), and yet if a Baron or 310 goes by, just close the window and go play their game. I also like the aerobatic capabilities. I myself don't care much for the hardcore yank-and-jerk stuff, but I do like the big and graceful loops and rolls (a la Hoover style). The Glasair does these maneuvers very nicely. I hope to have mine flying by this summer. Signed: Dale Truner, Captain, Republic Airlines."

As of this writing over 500 Glasair kits have been sold, and more than 50 are flying.

Specifications for the Glasair TD (tail dragger):

Dimensions
    Length: 18.7 ft.
    Height: 7.0 ft.
    Span: 23.3 ft.
    Wing area: 81.2 sq. ft.

Weights
    Empty: 925 lbs.
    Gross: 1500 lbs.
    Fuel cap: 42 gal.

Cabin
    Length: 45 in.
    Width: 39 in.
    Height: 37 in.
    Baggage: 80 lbs.

Performance
    Takeoff run   (solo):   390 ft.
                     (gross): 790 ft.
    Takeoff over
      50' obstacle   (solo):   755 ft.
                     (gross): 1425 ft.
    Climb   (solo): 1900 fpm.
                     (gross): 1300 fpm.
    Cruise   (75%):   224 mph.
                     (55%):   177 mph.

Fuel usage	(75%)	10.5 gph.
	(55%):	6.3 gph.
Stall	(flaps/solo):	59 mph.
	flaps/gross):	63 mph.
Service ceiling:		20,000 ft.
Landing roll	(solo):	475 ft.
	(gross):	550 ft.
Approach speed:		75 mph.
Landing speed:		65 mph.

Specifications for the Glasair RG (retractable):

Dimensions
Length:	18.7 ft.
Height:	6.5 ft.
Span:	23.3 ft.
Wing area:	81.2 sq. ft.

Weights
Empty:	1090 lbs.
Gross:	1800 lbs.
Fuel Cap:	42 gal.

Cabin
Length:	45 in.
Width:	39 in.
Height:	37 in.
Baggage:	80 lbs.

Performance
Takeoff run	(solo):	380 ft.
	(gross):	630 ft.
Takeoff over		
50' obstacle	(solo):	755 ft.
	(gross):	1200 ft.
Climb	(solo):	2300 fpm.
	(gross):	1400 fpm.
Cruise	(75%):	234 mph.
	(55%):	200 mph.
Fuel usage	(75%):	10 gph.
	(55%):	6.1 gph.
Range	(75%):	1150 mi.

Stall	(flaps/solo):	59 mph.
	(flaps/gross):	63 mph.
Service ceiling:		20,000 ft.
Landing roll	(solo):	435 ft.
	(gross):	530 ft.
Approach speed:		75 mph.
Landing speed:		65 mph.

Construction

The Glasair comes in the form of a kit. These kits are some of the most complete and comprehensive in the world. They include all parts necessary to build the airframe. Bolts, nuts, washers, rivets, bearings, bushings, pulleys, rod ends, hinge stock, and so forth are all included in the kits. Prop extensions, fuel systems, exhaust systems, prestamped parts, welded assemblies, and machined fittings are just some of the standard kit items. Even seat belts and shoulder harnesses with the Glasair logo embroidered on them are supplied. With all these items supplied in the kit, much of the hunting around for miscellaneous items is eliminated, saving a lot of time.

The typical building time is 1200 hours. This low building time is no doubt due to the extremely well engineered kit package, and its user-oriented instructions.

For further information contact:

Stoddard-Hamilton Aircraft Inc.
18701 58th Ave. N.E.
Arlington, WA 98223
Phone: (206) 435-8533

LANCER

Completely new to the market as a complete kit is the Lancer (Fig. 8-7). The Lancer, as described in the NEICO literature, is produced with the goal of being the most complete kit legally possible under current FAA regulations.

The Lancer is a two place, side-by-side, all-composite airplane with a top speed of 211 mph and range of better than 1000 miles. It has retractable

Fig. 8-7. Lancer. (courtesy NEICO)

tricycle landing gear and is powered by the Continental O-200 engine of Cessna 150 fame. The propeller is fixed-pitch wood.

Currently NEICO is preparing to test the 188-hp Lycoming O-235 engine, and the O-320 of 150 hp, on the Lancer.

The Lancer will be available as a kit starting in mid 1985.

Specifications for the Lancer 200 (O-200 engine):

Dimensions
    Length: 19.5 ft.
    Span: 23.4 ft.
    Wing area: 75.0 sq. ft.

Weights
    Empty: 650 lbs.
    Gross: 1275 lbs.
    Fuel Cap: 35 gal.
    Fuel usage: 5.5 to 6 gph.
    Cabin width: 42 in.

Performance
    Takeoff run   (solo):   400 ft.
    Rate of climb (solo):   1500 fpm.
    Cruise   (75%):   192 mph.
    Stall   58 mph.
    Landing roll   550 ft.

Construction

Twenty-seven premolded parts make up the entire airframe. The parts are made utilizing prepreg glass cloth and Nomex Honeycomb cores. All parts are vacuum-bagged and oven-cured, producing exceptionally light, strong parts. All mating surfaces are marked to ease accurate drilling and assembly. All hardware assemblies are precut and welded.

For further information contact:

NEICO
1015 W. 190th St.
Gardena, CA 90248
Phone: (213) 327-8851

POLLIWAGEN

The idea for the Polliwagen began in 1973, with testing via a quarter-scale radio-controlled model. The name Polliwagen comes from the craft's resemblance to a polliwog in shape, and the prototype having a Volkswagen engine.

In 1976 the first full-scale Polliwagen was displayed at Oshkosh. It is now another of the most successful composite homebuilt designs made.

The Polliwagen seats two side-by-side, and has retractable tricycle landing gear (Fig. 8-8). The plane can be built with a choice of three engines: the Volkswagen conversion, Continental O-200, or the Lycoming O-320. Naturally, the flight characteristics will vary depending upon individual engine selection.

The Polliwagen is a sophisticated aircraft, right down to the standard two-axis electric trim for roll and pitch correction. This is easily adaptable to autopilot usage.

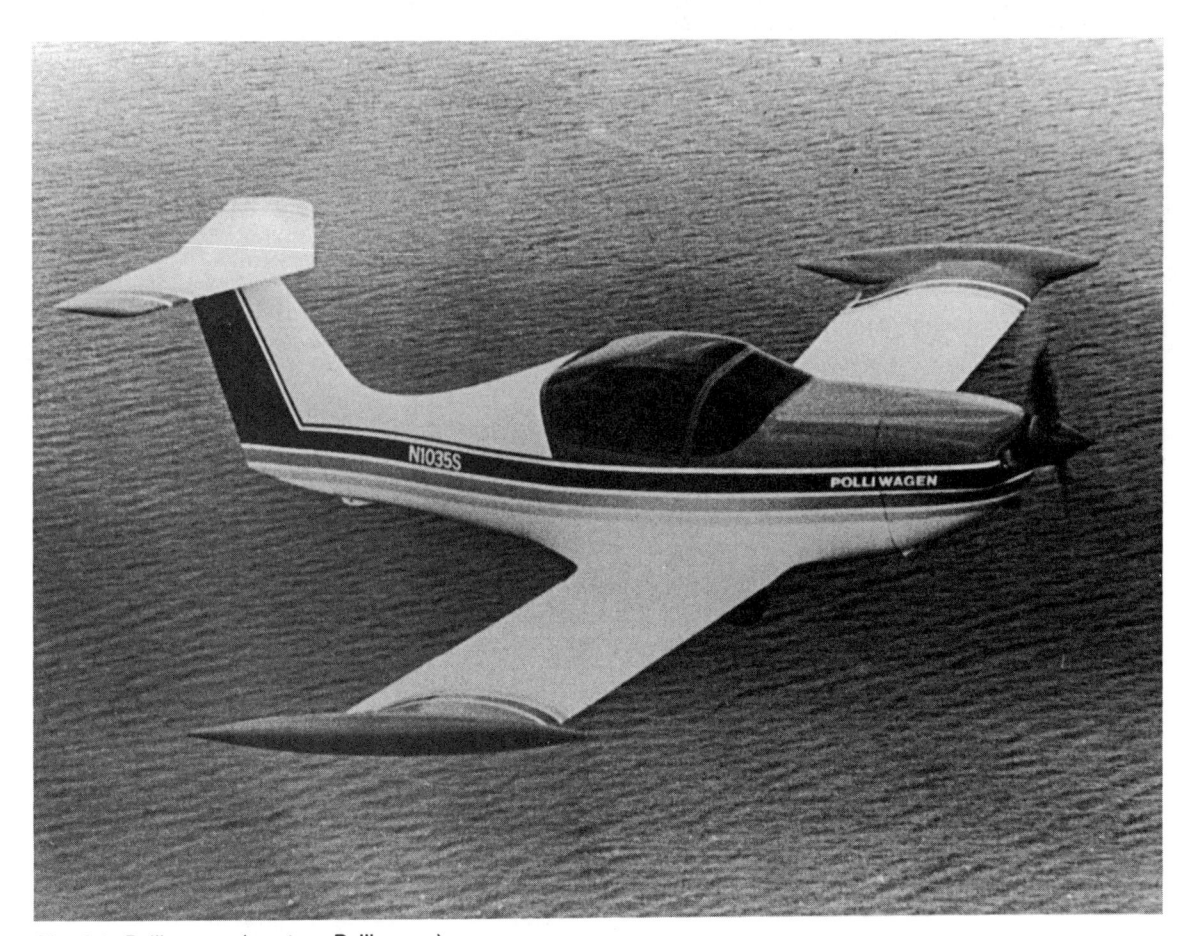

Fig. 8-8. Polliwagen. (courtesy Polliwagen)

The fiberglass composite material used in the Polliwagen is woven bidirectional fiberglass cloth saturated with epoxy resin in the layup process, then allowed to cure. Most of the craft's strength is obtained by combining a high-density fiberglass cloth inpregnated with epoxy resin with a lightweight low-density foam core—in a word, "composite."

The Polliwagen is trailerable by the removal of the wings. They attach to the airframe with two 3/4" 4130 steel pins. The electrical and fuel connections are made with quick disconnects.

Specifications for the Polliwagen:

Dimensions
  Length: 16.0 ft.
  Height: 5.6 ft.
  Span: 26.0 ft.
  Wing area: 90.0 sq. ft.

Weights
  VW Empty: 650 lbs.
  VW Gross: 1250 lbs.
  Cont empty: 650 lbs.
  Cont gross: 1350 lbs.
  Lyc empty: 750 lbs.
  Lyc gross: 1500 lbs.
  Cabin width: 40 in.

Performance
  Takeoff run   VW: 500 ft.
                Cont: 420 ft.
                Lyc: 360 ft.
  Climb (solo): VW: 900 fpm.
                Cont: 1500 fpm.
                Lyc: 2500 fpm.
        gross) VW: 700 fpm.
                Cont 1300 fpm.
                Lyc: 1500 fpm.
  Cruise (75%): VW: 160 mph.
                Cont: 190 mph.
                Lyc: 220 mph.
  Range (75%)   VW: 1520 mi.
                Cont: 1330 mi.
                Lyc: 1200 mi.
  Stall (flaps/solo) VW: 48 mph.
                Cont: 52 mph.
                Lyc: 57 mph.
      (flaps/gross) VW: 51 mph.
                Cont: 57 mph.
                Lync: 60 mph.
  Landing roll   475 ft.

Construction

The Polliwagen, like all other homebuilt aircraft, has "given" times for construction. However, remember these times are only for informational purposes, as they will vary from individual to individual. The claims are from 500 to 1500 hours, depending on the number of premolded parts used.

For further information contact:

Polliwagen
40940 Eleanora Way
Box 860
Murrieta, CA 92362
Phone: (714) 677-7877

QUICKIE

The Quickie products are among the lowest priced, most economical, lowest completion time homebuilt sport aircraft ever designed.

Based on the original single-place Quickie (Fig. 8-9), the Q-2 satisfies the many builders that require a two-place aircraft— with accompanying better performance figures (Fig. 8-10).

First flown on July 1, 1980, the Q-2 is powered with a Revmaster 2100-DG Volkswagen engine of 65 hp. The Q2 cockpit will accommodate a 6'8" 215 pound pilot plus a passenger, and has over 4 cu.ft. of baggage space behind the seats. The cabin width is greater than that of a Cessna 172.

The Q2 Turbo, using the Revmaster engine with the turbo option, is the same as the basic Q2 except for more power on demand from the engine, and higher cruising altitudes (Fig. 8-11).

For those pilot/builders wanting increased power and performance, Quickie has introduced the Q-200, which uses the Continental O-200 engine of 100 hp.

When it comes to strength, the Quickie excels. The main wing is stressed to 12 Gs and the canard, which is also the landing gear, is stressed to 36 Gs.

Quickie has an interesting claim in their literature that bears thinking about: They state that in 1983 they sold more two-place airplanes than Piper, Beech, and Cessna *combined*.

The Quickie Newsletter serves as a medium for

Fig. 8-9. Quickie. (courtesy Marion Pyles, Air-Pix Photos)

Fig. 8-10. Q-200. (courtesy Marion Pyles, Air-Pix Photos)

Fig. 8-11. Q-2 Turbo. (courtesy Marion Pyles, Air-Pix Photos)

getting plans updates and construction tips to the builders.

Specifications for the Quickie Q2:

Dimensions
    Length: 64 in.
    Width: 44 in.
    Height: 36 in.
    Baggage: 40 lbs.

Performance
    Takeoff run 610 ft.
    Climb: 1200 fpm.
    Cruise (75%): 170 mph.
    Fuel usage (75%): 42 mpg.
    Range (75%): 682 mi.
    Stall (power off): 60 mph.
           (power on): 58 mph.
    Service ceiling: 15,000 ft.
    Landing roll 950 ft.

Specifications for the Quickie Q2 Turbo:

Dimensions
    Length: 64 in.
    Width: 44 in.
    Height: 36 in.
    Baggage: 40 lbs.

Performance
    Takeoff run 525 ft.
    Climb: 1500 fpm.
    Max speed: 215 mph.
    Fuel usage (75%): 45 mpg.
    Stall (power off): 60 mph.
           (power on): 58 mph.
    Service ceiling: 30,000 ft.
    Landing roll 950 ft.

Specifications for the Quickie Q-200:

Dimensions
    Length: 19.8 in.
    Span: 16.8 ft.
    Wing area: 67.0 sq.ft.

Weights
  Empty: 505 lbs.
  Gross: 1100 lbs.
  Fuel cap: 20 gal.

Cabin
  Length: 64 in.
  Width: 44 in.
  Height: 36 in.
  Baggage: 40 lbs.

Performance
  Takeoff run   610 ft.
  Climb: 1600 fpm.
  Cruise (75%):   207 mph.
  Fuel usage (75%):   33 mpg.
  Range (75%):   530 mi.
  Stall   (power off):   64 mph.
                  (power on):   62 mph.
  Service ceiling: 21,000 ft.
  Landing roll   950 ft.

Construction

The Quickie is a kit consisting of everything except the battery and paint. Even included are the special tools you will need. No welding, machining, or searching for parts are required. Additionally, the hard-to-make parts are already made for you; these are the fuselage, cowling, and canopy (which comes premounted). There are other optional subassemblies that can be purchased to make the work go even quicker. The claimed construction times for a Q2 or Q-200 is 500 hours. This is considerably less than most other homebuilt airplanes.

For more information contact:

    Quickie Aircraft Corp.
    Box 786
    Mojave, CA 93501
    Phone: (805) 824-4313

RAND

Conceived by the late Ken Rand, the Rand airplanes are designed to provide good performance with VW conversion engines.

The KR-1 is a single-seat low-wing retractable monoplane which is powered by the smaller Volkswagen engines. It can be fitted with larger VW conversions up through the 2100cc engine (Fig. 8-12).

The KR-1B motorglider is a standard KR-1 with modified outer wing sections. With this simple wing conversion the KR-1 becomes a self-powered glider.

The KR-2 is the most popular Rand airplane. It is a two-seat (side-by-side) aircraft, with removable wings for ease of storage (Figs. 8-13, 8-14). It is powered with a converted Volkswagen engine. Any VW conversion from 1600cc to 2100cc will do the job. Naturally, the more power, the better the performance.

The cost and construction time required for completion are minimized by the use of a combination of wood construction and composite construction.

Construction of the typical Rand fuselage is similar to building a balsa wood model airplane, just on a larger scale. It is a foam-over-structure composite, having a wooden structure built to carry the loads. The foam is added to give smooth contour and shape. This foam is covered with fiberglass. This is less scary to the first-time builder than a more complex, completely composite structure.

A newsletter for KR pilots and builders is available from:

    KR Newsletter
    Box 4113
    Elglewood, CO 80155

Specifications for the KR-1:

Dimensions
  Length: 12.8 in.
  Wing span: 17.0 ft.
  Wing area: 62.0 sq.ft.

Weights
  Empty weight: 375 lbs.
  Gross: weight: 750 lbs.
  Baggage: 20 lbs.

Performance
  Takeoff dist: 350 ft.
  Landing dist: 900 ft.
  Stall: 52 mph.

Fig. 8-12. KR-1. (courtesy Marion Pyles, Air-Pix Photos)

Fig. 8-13. KR-2. (courtesy Marion Pyles, Air-Pix Photos)

Fig. 8-14. KR-2 with unusual canopy similar to a Swift.

Max speed:	200 mph.
Cruise speed::	180 mph.
Range:	1400 mi.
Climb (gross):	800 fpm.
Svc ceiling:	15,000 ft.
Engine:	1834cc VW.
Fuel usage:	3.8 gph.

Specifications for the KR-1B:

Dimensions
Length:	12.9 in.
Wing span:	27.0 ft.
Wing area:	91.0 sq.ft

Weights
Empty:	484 lbs.
Gross:	800 lbs.
Baggage:	20 lbs.

Performance
Takeoff dist:	300 ft.
Landing dist:	300 ft.
Stall:	38 mph.
Max speed:	144 mph.
Glide ratio:	20 to 1
Cruise speed:	130 mph.

Specifications for the KR-2:

Dimensions
Length:	14.5 in.
Wing span:	20.8 ft.
Wing area:	80.0 sq.ft.

Weights
Empty weight:	480 lbs.
Gross weight:	900 lbs.
Baggage:	35 lbs.

Performance
Takeoff dist:	350 ft.
Landing dist:	900 ft.
Stall:	52 mph.

Max speed:	200 mph.
Cruise speed:	180 mph.
Range:	1600 mi.
Climb (gross):	1200 fpm.
Svc ceiling:	15,000 ft.
Engine:	2100cc VW
Fuel usage:	3.8 gph.

Construction

The KR series airplanes come as complete kits, containing all you need for completion except the engine. Many parts are premolded, and all welding and some machining is done for you. There are completed subassemblies available from Rand (and other companies) that will speed construction.

For further information contact:

Rand/Robinson Engineering, Inc.
5842 K McFadden Ave.
Huntington Beach
CA 92648
Phone: (714) 898-3811

RUTAN

The Rutan Aircraft Factory (RAF) was formed in 1969. Its purpose was to research nonconventional aircraft. Since that time homebuilding has not been the same. The modern era of composite construction is often referred to as the Rutan Revolution for two reasons—the unique canard designs, and the approach to composite construction of aircraft.

Burt Rutan, through RAF, has introduced several composite aircraft designed for the homebuilder. All have been very successful airplanes, with many built, or being built. For a complete history of Rutan and his airplanes, I suggest reading *The Complete Guide to Rutan Aircraft* by Don and Julia Downie (TAB #2360).

The first composite airplane introduced to the homebuilder by RAF was the famous VariEze, a small two-place canard-type aircraft (Fig. 8-15). Its structure lead the way for moldless composite airplane construction. The prototype VariEze was built in early 1975, and first flown in May of that year. This history-making airplane is now on display at the EAA Air Museum.

The following year a revised version of the VariEze was built and flown. It is designed around the O-200 Continental Teledyne engine, and is a somewhat heavier aircraft. RAF states, "Approximately 400 VariEzes have been completed by homebuilders worldwide. They have logged over 150,000 flight hours, which represents the most experience of any powered all-composite aircraft type."

In June 1978 the Defiant was first flown. It is a four-place centerline-thrust twin of canard design (Fig. 8-16). It is a very basic design, having no flaps, retractable landing gear, hydraulic system, oleo struts, constant-speed prop, etc. The lack of these systems is a real plus for the Defiant—if it's not there, it can't break. Being centerline thrust, the pilot's work is greatly simplified in the event of an engine failure. For example, in the event of engine failure on takeoff, only airspeed must be maintained. There is no need to immediately identify the dead engine, or to correct trim. Just maintain airspeed and continue to climb. This is a real safety feature. The Defiant has only recently been made available to the homebuilder.

In 1979 RAF built and tested the Long-EZ. Basically, it is an enlarged VariEze, offering greater useful load and improved range. Additionally, the flight characteristics are more docile than its predecessor. The Long-EZ was designed to be very safe: free of stalls and departure high-angle-of-attack characteristics. This was proven by NASA flight testing in 1981. RAF states, "Our Long-EZ is so efficient, the engine can be shut down while at 5 foot altitude over the numbers at only 120 knots, then it can pull up, fly a 360 degree pattern, and land on the same runway—completely without power!" The aircraft set a long-distance record of 4800 miles in December 1979. The Long-EZ has become the most popular RAF airplane.

The Solitaire, made available to the homebuilders in 1983, was first designed and built for entry into a Soaring Society of America competition involving the design of self-launchable gliders, which it won. The Solitaire is a canard glider with a glide ratio of 32:1 (Fig. 8-17). However, the best feature is the retractable KFM

Fig. 8-15. VariEze and Long-EZ. (courtesy RAF)

engine, which makes the craft self-launchable. Goodbye towing fees! The engine is deployable in the air and has electric start. Just imagine the limitless flight possible with this configuration. Unlike the other RAF airplanes, the Solitaire construction is not all moldless. The fuselage comes as two premolded pieces, and most other sections are either included in the kit prefabricated or premachined. Only the wings are built in moldless fashion. RAF states, "We estimate that an average builder, purchasing all the available parts, could build the aircraft in 400 hours at a cost of between $7000 and $9000."

Specifications for the Defiant:
Engine: (2) 16-hp O-320 Continental
Dimensions
  Span: 31.4 ft.
  Wing area: 133 sq.ft.

Weights
  Empty: 1600 lbs.
  Gross: 2950 lbs.
  Fuel cap: 115 gal.

Performance (at gross weight)
  Takeoff run: 1480 ft.
  Climb: single   1500 fpm.
    engine climb: 310 fpm.
  Cruise   (normal
                  70%):   184 kts.
                  (economy
                  55%):   168 kts.
  Fuel usage   (normal):   17.8 gph.
                  (economy):   13.9 gph.
  Range   (normal):   1044 nm.
                  (economy):   1208 nm.
  Stall: 64 kts.

Fig. 8-16. Defiant. (courtesy RAF)

Specifications for the Long-EZ:
Engine: 108-hp O-235 Lycoming

Dimensions
  Span: 26.1 ft.
  Wing area: 94.8 sq.ft.

Weights
  Empty: 800 lbs.
  Gross: 1425 lbs.
  Fuel cap: 52 gal.

Performance
  Takeoff run: (at gross): 950 ft.
  Climb: 1250 fpm.
  Cruise (normal 75%): 186 kts.
  (economy 40%): 146 kts.
  Fuel usage (normal): 6.6 gph.
  (economy): 3.6 gph.

Range (normal): 1150 nm.
  (economy): 1690 nm.
Landing roll: 680 ft.

Specifications for the Solitaire:
Engine: 23-hp KFM 107E

Dimensions
  Span: 41.75 ft.
  Wing area: 102.5 sq.ft.

Weights
  Empty: 380 lbs.
  Gross: 620 lbs.
  Fuel cap: 5 gal.

Performance
  L/D at 50 kts.: 32: 1
  Minimum sink: 150 ft./40 kts.
  Minimum flying speed: 32 kts.
  V_{ne}: 115 kts.

Construction

The RAF sells plans only; however, they are supported by two suppliers of complete kits of raw materials and prefabricated parts:

>Aircraft Spruce and Specialty Co.
>201 W. Truslow
>Box 424
>Fullerton, CA 92643
>Phone: (714) 870-7551

>Wicks Aircraft Supply Co.
>401 Pine St.
>Highland, IL 62249
>Phone: (618) 654-7447

Prefabricated machined parts such as control systems, welded systems, exhaust systems, etc., are available from:

>Ken Brock Manufacturing Co.
>11852 Western Ave.
>Stanton, CA 90680
>Phone: (714) 898-4366

Many premade subassemblies are available for RAF airplanes from other sources. Advertisements for such will be seen in the various aviation publications.

For further information about RAF airplanes contact:

>RAF
>Building 13
>Mojave Airport
>Mojave, CA 93501
>Phone: (805) 824-2645

Fig. 8-17. Solitaire. (courtesy RAF)

SEA HAWK

During his years in his native Canada, Gerry LeGare saw the vital need for a superior amphibious aircraft. His revolutionary new Sea Hawk amphibian is now available to fill that requirement.

The Sea Hawk is a two-place plane that can be powered with any engine in the 70 to 160-hp range. This gives the builder a choice in economy and/or performance. The low landing and takeoff speeds, good freeboard, structural integrity, and corrosion resistance of the composite structure make the Sea Hawk an ideal candidate for salt-water usage (Fig. 8-18).

The Sea Hawk is quite at home on land and has retractable tricycle landing gear. The nosewheel swivels; therefore, differential braking is used for low-speed ground handling. If rough-field operation is being considered, it is recommended that the rough-field landing gear kit be purchased; this is called the Bush Hawk option. Unlike most pusher airplanes, there is no fear of propeller damage from picked-up debris, as the wings shield the prop.

Due to the baggage area being located at the CG point, there is no limitation to weight, save for overgrossing the airplane.

A newsletter, *Hawk Talk*, is published quarterly and contains news chatter, plans changes (clarifications), and building tips.

Construction

Building time is low, as the entire composite structure is provided to the builder in premolded form.

The fuselage comes with a premounted canopy, the vertical fin, and aft canopy bulkhead installed. The wings are also premolded, utilizing carbon fiber materials. The wings can be used as fuel tanks.

Completion time should be about 1000 hours. For further information contact:

Leg-Air Corp.
Building A2
Medford Airport
Medford, OR 97501
Phone: (503) 779-1207

SILHOUETTE

The Silhouette is an all-composite, predominantly premolded homebuilt capable of outperforming many factory aircraft of considerably higher power ratings. It is a single-seat, super-economical craft capable of providing over 50 mpg at a speed in excess of 110 mph (Fig. 8-19). It is trailerable, having removable wings, and can even be picked up at the factory in kit form on the trailer you will later use to transport it to the airport.

Specifications of the Sea Hawk (equipped with a 150-hp Lycoming O-320-C engine):

Dimensions
  Length: 21.0 ft.
  Height: 7.5 ft.
  Span: 24.0 ft.
  Wing area: 118 sq.ft.
  Cabin Width: 44 in.

Weights
  Empty: 850 lbs.
  Gross: 1600 lbs.
  Fuel cap: 70 gal.

Performance
  Takeoff run: 500 ft.
  Climb: 1200 fpm.
  Cruise: 150 mph.
  Fuel usage: 7 gph.
  Range: 1320 mi.
  Stall: 42 mph.
  Service ceiling: 18,000 ft.
  Landing roll: 600 ft.

Specifications for the Silhouette:

Dimensions
  Length: 19.25 ft.
  Span: 31.2 ft.
  Wing area: 75 sq.ft.

Weights
  Empty: 430 lbs.
  Gross: 700 lbs.
  Fuel cap: 9.5 gal.
  Engine: Rotax 447 (44 hp).

Performance
  Takeoff run: 800 ft.
  Landing roll: 1000 ft.
  Stall: 48 mph.
  Cruise: 120 mph.
  Climb: 750 fpm.
  Endurance: 3 hrs.

(All numbers based on flight test from N84TR.)

Fig. 8-18. Sea Hawk. (courtesy Leg-Aire)

Construction

As a kit airplane, the Silhouette is designed to provide quick and easy satisfaction. No exotic tools are required for completion, and all parts are alignment marked. The following parts come premolded:

- ☐ Fuselage halves.
- ☐ Bulkheads.
- ☐ Consoles.
- ☐ Instrument panel.
- ☐ Wingtips.
- ☐ Wing root fairings.
- ☐ Nose bowl.
- ☐ Cowling covers.
- ☐ Landing gear.
- ☐ Turtledeck.
- ☐ Premounted canopy.
- ☐ Wing spars.

Precut foam cores are provided for the wing and tail assemblies. This negates the need for "hot wiring." Even paint is included in the kit. The claimed completion time is "much less than 500 hours."

For further information, contact:

Silhouette Aircraft
848 East Santa Maria St.
Santa Paula Airport
Santa Paula, CA 93060
Phone: (805) 525-4445

SOLO

Although not a homebuilt airplane, the Solo is included here as it does demonstrate the widespread application of composite structures in the field of aviation. The Solo is an Ultralight, falling within the guidelines of FAR 103.7 (Fig. 8-20).

Fig. 8-19. Silhouette. (courtesy Silhouette Aircraft Co.)

The airplane consists of fiberglass, Kevlar, graphite, and foam. The fuselage is constructed of Kevlar over foam for great strength. The wings are stressed to +6 Gs and −4 Gs. They too are composite in construction, with a main spar of graphite and foam, capped with graphite strips. Additionally, they fold for road transport.

Specifications for the Solo:

Dimensions
  Length: 19.5 ft.
  Height: 7.5 ft.
  Span: 30.0 ft.
  Wing area: 128.0 sq.ft.

Weights
  Empty: 275 lbs.
  Gross: 500 lbs.

Performance
  Takeoff run: 90 ft.
  Climb: 550 fpm.
  Cruise: 40-60 mph.
  Stall: 27 mph.
  Landing roll: 60 ft.

For further information about this ready-to-fly composite ultralight, contact:

American Air Technology
1290 Bodega Ave.
Petaluma, CA 94952
Phone: (707) 762-1800

STAR-LITE

The Star-Lite (Fig. 8-21) won the Outstanding

New Design award at Oshkosh in 1983. At the same time it also won first place in the ARV Design Competition. ARV stands for Air Recreational Vehicle, which is the current name given to the class of airplanes that offer extremely low cost flying.

The Star-Lite was designed for the specific purpose of lowering the cost of sport aviation. It is a conventional-gear airplane powered by the Rotax 447 geared engine.

The Rotax engine is a 40-hp two-cycle engine designed for airplane use. It has been proven in ultralight service, and is found on other homebuilt designs such as the Avid Flyer (a tube-and-fabric STOL two-place homebuilt).

The Star-Lite is completely trailerable, and can be unloaded and assembled in ten minutes.

Satisfying the FAA 51 percent rule, the Star-Lite comes as a kit, not just raw materials. The fuselage is vacuum-bag formed at the factory, while the wings are built of plywood with precut ribs.

Specifications for the Star-Lite:

Dimensions
- Length: 16.4 ft.
- Span: 21.5 ft.
- Wing area: 57.0 sq.ft.
- Height: 4.0 ft.

Weights
- Empty: 225 lbs.
- Gross: 450 lbs.
- Fuel cap: 5 gal.

Performance
- Climb: 800 fpm.
- Cruise: 110 mph.
- Range: 300 mi.
- Stall: 42 mph.

Construction

Every part of the Star-Lite aircraft that requires a specialized skill to build is supplied in the kit. Consequently, the construction time is reduced to about 400 hours.

For further information, contact:

Star-Lite Aircraft, Inc.
2219 Orange Blossom
San Antonio, TX 78247

Fig. 8-20. Solo. (courtesy American Air Technology)

Fig. 8-21. Star-Lite. (courtesy Star-Lite Aircraft)

W.A.R. REPLICAS

Many a flier has had dreams of flying a World War II fighter plane, but, alas, like most of us poor folk, that dream will never be attained—that is, unless one can settle for a replica of one of the great fighters.

W.A.R. Replicas produces plans for 1/2-scale replicas of the German Focke-Wulf 190, F4U Corsair, and the P-47 Thunderbolt.

Each type is built on the same basic airframe; in a manner similar to the Rand KR airplanes, a wood airframe bears the flight loads. Foam and fiberglass are applied to this for contouring and looks.

The replicas are single-seat and are powered with the Continental O-200 engine. Three and four-bladed props, specially shaped cowlings, and properly scaled canopies are available for these little fighters.

Specifications for the W.A.R. Replicas:

Dimensions
   Length: 16.6 ft.
   Span: 20.0 ft.
   Wing area: 70.0 sq.ft.

Weights
   Empty: 670 lbs.
   Gross: 900 lbs.
   Fuel cap: 12 gal.

Performance
   Climb: 1000 fpm.
   Cruise: 145 mph.
   Range: 400 mi.
   Stall: 78 mph.

Construction

Materials kits are available from suppliers, much the same as Long-EZ kits are available.

For further information, contact:

W.A.R. Replicas
348 S. 8th St.
Santa Paula, CA 93060

Chapter 9

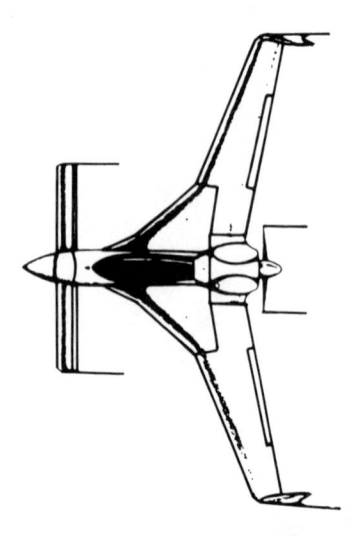

FAA Relations

The Federal Aviation Agency (FAA) should be viewed by the homebuilder—or any other member of small general aviation—as a friend.

The local FAA office—GADO, FSDO, MIDO—is where you go for help and guidance. Whether your problem is about homebuilts, policy, or even a consumer problem, the men and women of the local FAA office stand ready to help you. You ask the questions and they give the answers.

If you think I am impressed with the FAA, you are right. I have spent years visiting all kinds of government offices. I have seen wastes of money, talent, and physical space. Unfortunately, waste seems to be the government way.

But when I enter the local FAA office, there is calm and efficiency. You don't see "civil servants" sitting all over and around their desks happily chatting away about anything but work. Instead, you are efficiently asked what your needs are, and are taken to a person knowledgeable in that area.

Visit an FAA office, you might like to see what your tax money is doing. And by all means visit your local office *before* you spend the first dime towards an amateur-built aircraft. This one visit could save you thousands of dollars and hundreds of misspent hours of labor.

The FAA, in answer to the many questions about homebuilding, has issued Advisory Circular #20-27C. It is reproduced here:

CERTIFICATION AND OPERATION OF AMATEUR-BUILT AIRCRAFT

1. PURPOSE. This advisory circular (AC) provides guidance and information relative to the airworthiness certification and operation of amateur-built aircraft.

2. CANCELLATIONS.

   a. AC 27-27B, Certification and Operation of Amateur-Built Aircraft, dated April 20, 1972.

   b. AC 20-28A, Nationally advertised Construction Kits Amateur- Built Aircraft, dated December 29, 1972.

3. BACKGROUND.

   a. The FAA has received many requests from amateur aircraft builders for information relative to the construction, certification, and operation of amateur-built aircraft. This AC provides guidance concerning the building, certification, and operation of amateur-built aircraft of all types; defines the eligibility of nationally advertised kits; defines how much construction the amateur builder must do to have the aircraft eligible for airworthiness certification; and describes the FAA role in the certification process.

b. The Federal Aviation Regulations (FAR) provide for the issuance of experimental certificates to permit the operation of amateur-built aircraft. FAR 21.191(g) defines an amateur-built aircraft as an aircraft, the major portion of which has been fabricated and assembled by person(s) who undertook the construction project solely for their education or recreation. The FAA has interpreted this rule to require that more than 50 percent of the aircraft must have been fabricated and assembled by the individual or group. Commercially-produced components and parts which are normally purchased for use in aircraft may be used, including engines and engine accessories, propellers, tires, spring steel landing gear, main and tail rotor blades, rotor hubs, wheel and brake assemblies, forgings, castings, extrusions, and standard aircraft hardware such as pulleys, bellcranks, rod ends, bearings, bolts, rivets, etc.

4. DEFINITION. As used herein, the term "District Office" means the FAA General Aviation (GADO), Flight Standards (FSDO), or Manufacturing Inspection (MIDO) District Office that will perform the airworthiness inspection and certification of an amateur-built aircraft.

5. FAA INSPECTION CRITERIA.

a. In the past, the FAA has inspected amateur-built aircraft at several stages during the construction of the aircraft. These inspections are commonly referred to as precover inspections. The FAA also inspected the aircraft upon completion (prior to the initial issuance of the special airworthiness certificate necessary to show compliance with FAR 91.42(b)), and again prior to the issuance of the unlimited special airworthiness certificate. In the interest of streamlining operations within the government, and utilizing FAA inspection resources in areas most affecting safety, the FAA has reassessed its position concerning the need for all of these inspections. The following reflects the results of this assessment:

(1) The amateur-built program was designed to permit any person to build and operate an aircraft solely for educational and recreational purposes. The FAA has always maintained that the amateur builder may select his/her own design and should not be inhibited by any overly stringent FAA requirements. The FAA does not approve these designs since it would not be practical to develop design standards for the multitude of unique design configurations that are generated by amateur builders.

(2) FAA inspections of these aircraft are limited to ensuring the use of acceptable workmanship methods, techniques, and practices; verification of flight tests to ensure that the particular aircraft is controllable throughout its normal range of speeds and throughout all the maneuvers to be executed; and determination of operating limitations necessary to protect persons and property not involved in this activity.

(3) In recent years, amateur builders have adopted a practice whereby they call upon a person having expertise with aircraft construction techniques (such as Experimental Aircraft Association (EAA) designees, reference paragraph 7d(1)), to inspect particular components, e.g., wing assemblies, fuselages, etc., prior to covering, and conduct other inspections, as deemed necessary. This practice has been highly successful in ensuring construction integrity.

(4) There are many instances where the FAA has found that precover inspections were unnecessary, since in some cases the areas requiring inspection are readily accessible when the aircraft was completed. In other instances, precover inspections were found to be neither meaningful nor feasible, such as in cases involving aircraft constructed from composite materials.

(5) The FAA inspection previously performed after successful completion of the flight test, and prior to issuance of an unlimited certificate, was determined to be redundant in that any workman-

ship discrepancies would be detected during inspections performed prior to the issuance of the initial special airworthiness certificate.

b. In view of the foregoing considerations, the FAA has concluded that safety objectives, relative to the amateur-built program, can be best accomplished, with no derogation of safety, by the use of the following criteria:

(1) Amateur builders should continue the practice of having knowledgeable persons (i.e., EAA designees, FAA certified mechanics, etc.) perform precover inspections and other inspections as appropriate. In addition, builders should document construction using photographs taken at appropriate times prior to covering. The photographs should clearly show methods of construction and quality of workmanship. Such photographic records should be included with the builder's log and other construction records.

(2) The FAA will conduct an inspection of the aircraft prior to the issuance of the initial special airworthiness certificate to enable the applicant to show compliance with FAR 91.42(b). This inspection will include a review of the information required by FAR 21.193, the aircraft builder's logbooks, and an examination of the completed aircraft to ensure that proper workmanship has been used in the construction of the aircraft. Appropriate operating limitations will also be prescribed at this time.

(3) Upon completion of the required flight test, the FAA will review the results of the tests accomplished, as recorded in the aircraft logbook. Upon a determination that compliance has been shown with FAR 91.42(b), the FAA will issue the unlimited special airworthiness certificate with expanded operating limitations.

6. CERTIFICATION STEPS. The following procedures are in the general order to be followed in the certification process:

a. Initial Step. The prospective builder should contact the nearest District Office to discuss his plans for building the aircraft with an FAA inspector. During this contact, the type of aircraft, its complexity and/or materials should be discussed. The FAA will provide the prospective builder with any guidance necessary to ensure a better understanding of applicable regulations.

b. Registration. Prior to completion of the aircraft, the builder may apply for an identification number and register the aircraft. Detailed procedures are in paragraph 7A. This must be done before submitting an Application for Airworthiness Certificate, FAA Form 8130-6, under FAR 21.173.

c. Marking. The identification number (N-number) assigned to the aircraft, and an identification plate must be affixed in accordance with FAR 21.182 and FAR Part 45, Identification and Registration Marking. Detailed procedures are in paragraph 7b.

d. Application. The builder may apply for an experimental certificate by submitting the following documents and data to the nearest District Office:

(1) Application for Airworthiness Certificate, FAA Form 8130-6.

(2) Enough data (such as photographs) to identify the aircraft.

(3) An Aircraft Registration Certificate, AC Form 8030-3, or the pink copy of the Aircraft Registration Application, AC 8050-1.

(4) A statement setting forth the purpose for which the aircraft is to be used; i.e. "operating an amateur-built aircraft, FAR 21.191(g)."

(5) A notarized statement that the applicant fabricated and assembled the major portion (reference paragraph 7e) of the aircraft, for education or recreation, and has evidence to support the statement available to the FAA upon request.

(6) Evidence of inspections, such as a logbook entry signed by the builder describing all inspections conducted during construction of the aircraft. This will substantiate that the construction has been accomplished in accordance with acceptable workmanship methods, techniques, and practices, in addition to photographic documentation of construction details.

e. FAA Inspection and Issuance of Airworthiness Certificate.

(1) After inspection of the documents and data submitted with the application, the FAA will inspect the aircraft, and upon a determination that

the aircraft has been properly constructed will issue an experimental certificate together with operating limitations that specify the flight test area. The inspector will verify that all required markings are properly applied, including the following placard which must be displayed in the cabin or the cockpit at a location where it is in full view of all the occupants:

PASSENGER WARNING—THIS AIRCRAFT IS AMATEUR-BUILT AND DOES NOT COMPLY WITH FEDERAL SAFETY REGULATIONS FOR STANDARD AIRCRAFT

(2) Details concerning flight test areas are contained in paragraph 7c. The operating limitations are a part of the experimental certificate and must be displayed with the certificate when the aircraft is operated. It is the responsibility of the pilot to conduct all flights in accordance with the operating limitations, as well as the General Operating and Flight Rules in FAR Part 91.

(3) Upon satisfactory completion of operations in the assigned test area, the owner of the aircraft may apply to the FAA for amended operating limitations by submitting to the nearest District Office an Application for Airworthiness Certificate, FAA Form 8130-6. Prior to issuance of the amended limitations, the FAA inspector will review the applicant's flight log to determine whether corrective actions have been taken on any problems encountered during the testing and that aircraft condition for safe operation has been established.

7. GENERAL INFORMATION

a. Aircraft Registration. The FAR requires that all U.S. civil aircraft be registered before an airworthiness certificate can be issued. FAR Part 47, Aircraft Registration, prescribes the requirements for registering civil aircraft. The basic procedure is as follows:

(1) Before an amateur builder can register his aircraft he must first obtain an identification number which will eventually be displayed on the aircraft. It is not necessary to obtain an identification number at an early stage in the project, especially if the builder intends to obtain a special number of his choice. Under FAR Part 47, a special identification number costs $10.00 and may be reserved for no longer than one year. Although this reservation does not apply to numbers assigned at random by the Aircraft Registry, the best time to request identification number assignments in either case is in the later stages of construction.

(2) To apply for either a random identification number or a special identification number, the owner of an amateur-built aircraft must provide information required by the FAA Aircraft Registry by completing an Affidavit of Ownership for Amateur-Built Aircraft, AC Form 8050-88. The affidavit establishes the ownership of the aircraft. If the aircraft is built from a kit, the builder should also submit a signed bill of sale from the manufacturer of the kit as evidence of ownership. Any communication sent to the FAA concerning aircraft registration should be mailed to the FAA Aircraft Registry, Department of Transportation, P.O. Box 25504, Oklahoma City, Oklahoma 73125.

(3) After receipt of the applicant's letter requesting identification number assignment, the FAA Aircraft Registry will send a form letter giving the number assignment, a blank Aircraft Registration Application, AC Form 8050-1, and other registration information. All instructions must be carefully complied with to prevent return of the application and delay in the registration process. The Aircraft Registration Application white original and green copy should be returned to the FAA Aircraft Registry within 90 days prior to the estimated completion date of the amateur-built aircraft, accompanied by a fee of $5.00 payable by check or money order to the Federal Aviation Administration. The pink copy of the application should be retained by the applicant to be carried in the aircraft as a temporary registration until the Certificate of Registration is issued.

b. Identification Marking. When a builder applies for an airworthiness certificate of his amateur-built aircraft, he must show in accordance with FAR 21.182 that his aircraft bears identification and registration markings required by FAR Part

45. The following summary of the pertinent rules is provided for guidance.

(1) The aircraft must be identified by some means of a fireproof plate that is etched, stamped, engraved, or marked by some other approved fireproof marking as required by FAR 45.11. The identification plate must include the information required by FAR 45.13.

(2) The identification plate must be secured in such a manner that it will not likely be defaced or removed during normal service, or lost or destroyed in an accident. It must be secured to the aircraft at an accessible location near an entrance, except that if it is legible to a person on the ground it may be located externally on the fuselage near the tail surfaces.

(3) The builder's name on the data plate must be the amateur builder, not the designer, plans producer, or kit manufacturer. The model is whatever the builder wishes to assign, provided it will not be confused with other commercially built aircraft model designations.

(4) The required registration marks on most amateur-built airplanes must be displayed on either the vertical tail surfaces or the sides of the fuselage. However, the builder should refer to FAR 45.25, which defines specific requirements for the location of marks on fixed-wing aircraft. The location of registration marks for rotorcraft, airships, and balloons is specified in FAR 45.27. These registration marks must be painted on or affixed by any other means ensuring a similar degree of permanence. Decals are also acceptable. The use of tape which can be peeled off or water-soluble paint, such as poster paint, is not considered acceptable.

(5) Most fixed-wing amateur-built aircraft are eligible to display registration marks with a minimum height of three (3) inches. However, if the maximum cruising speed of the aircraft exceeds 180 knots (207 miles per hour), registration marks at least twelve (12) inches high must be displayed. The builder should refer to FAR 45.29, which defines the minimum size and proportions for registration marks on all types of aircraft.

(6) The identification marks displayed on the aircraft must consist of the Roman capital letter "N" (denoting United States registration) followed by the registration number of the aircraft. Any suffix letter used in the marks must also be a Roman capital letter. In addition, the word "experimental" must also be displayed on the aircraft near each entrance to the cabin or cockpit in letters not less than 2 inches nor more than 6 inches in height.

(7) If the configuration of the aircraft prevents the aircraft from being marked in compliance with any of the above requirements, the builder should contact the FAA regarding approval of a different marking procedure. It is highly recommended that any questions regarding registration marking be discussed and resolved with the local FAA Inspector before the marks are affixed to the aircraft.

c. Flight Test Areas.

(1) Amateur-built airplanes and rotorcraft will initially be limited to operation within an assigned flight test area for at least 25 hours when a type certificated (FAA-approved) engine/propeller combination is installed, or 40 hours when an uncertified (not FAA-approved) engine/propeller combination is installed. Amateur-built gliders, balloons, dirigibles and ultralight vehicles built from kits evaluated by the FAA and found eligible to meet requirements of FAR 21.191(g), for which original airworthiness certification is sought will be limited to operation within an assigned flight test area for at least 10 hours of satisfactory operation, including at least five takeoffs and landings.

(2) The desired flight test area should be requested by the applicant, and if found acceptable by the FAA Inspector will be approved and so specified in the Operating Limitations. It will usually encompass the area within a twenty-five (25) statute mile radius from the aircraft's home base. The FAA will ensure that the area selected is not over densely populated areas or in congested airways so that the flight testing, during which passengers may not be carried, would not likely impose any hazard to persons or property not involved in this activity. The shape of the flight test area selected may need to modified to satisfy these requirements.

(3) The carrying of passengers or other crew

members will not be permitted unless they are necessary to the conduct of the flight test while the aircraft is restricted to the flight test area.

(4) When it is shown in accordance with FAR 91.42(b) that the aircraft is controllable throughout its normal range of speeds and all maneuvers to be executed, and has no hazardous operating characteristics or design features, and the time period in the flight test area has been completed, the owner may apply for operation outside the assigned flight test area.

d. Design and Construction.

(1) Many individuals who desire to build their own aircraft have little or no experience with respect to aeronautical practices, workmanship, or design. An excellent source of advice in such matters is the Experimental Aircraft Association (EAA) located in Oshkosh, Wisconsin. The EAA is an organization established for the purpose of promoting amateur building and giving technical advice and assistance to its members. The EAA has implemented a designee program, whose aim is to ensure the safety and dependability of amateur-built aircraft. Most EAA designees are willing to inspect amateur-built aircraft projects and offer constructive advice regarding workmanship and/or design. The FAA strongly encourages the use of such designees.

(2) Any choice of engines, propellers, wheels, and other components, and any choice of materials may be used in the construction of an an amateur-built aircraft. However, it is recommended that FAA-approved components and established aircraft quality material be used, especially in fabricating parts constituting the primary structure, such as wing spars, critical attachment fittings, and fuselage structural members. Inferior materials, or materials whose identity cannot be established, should not be used.

(3) The design of the cockpit or cabin of the aircraft should avoid, or provide for padding on, sharp corners or edges, protrusions, knobs and similar objects which may cause injury to the pilot or passengers in the event of an accident.

(4) An engine installation should be such that adequate fuel is supplied to the engine in all anticipated flight attitudes. Also, a suitable means, consistent with the size and complexity of the aircraft, should be provided to reduce fire hazards wherever possible, including a firewall between the engine compartment and the fuselage. When applicable, a system providing for carburetor heat should also be provided to minimize the possibility of carburetor icing.

(5) Much additional information and guidance concerning acceptable fabrication and assembly methods, techniques, and practices are provided in FAA Advisory Circular (AC) No. 43.13-1A "Acceptable Methods, Techniques, and Practices—Aircraft Inspection and Repair," and AC No. 43-13-2A, "Acceptable Methods, Techniques and Practices—Aircraft Alterations." These publications are available from the Government Printing Office.

(6) In the areas of engineering design, the builder should obtain the services of a qualified aeronautical engineer, or consult the seller of purchased plans or construction kits, as appropriate.

e. Construction Kits.

(1) Construction kits containing raw materials and some prefabricated components may be used in building an amateur-built aircraft; however, aircraft which are assembled from kits composed of completely finished prefabricated components and parts, and precut, predrilled materials are not considered to be eligible for certification as amateur-built aircraft, since the major portion of the aircraft would not have been "fabricated and assembled" by the builder.

(2) All aircraft built from a kit may be eligible for certification, provided that the major portion (more than 50 percent) has been "fabricated and assembled" by the amateur builder. The major portion of such a kit may consist of raw stock such as lengths of wood, tubing, extrusions, etc., which may have been cut to an appropriate length. A certain quantity of prefabricated parts such as heat-treated ribs, bulkheads or complex forms made from sheet metal, fiberglass, or polystyrene would also be acceptable, provided it still meets the major portion of "fabrication and assembly" requirement and the amateur builder satisfies the

FAA Inspector that completion of the aircraft is not merely an assembly operation.

(3) Various material/part kits for the construction of aircraft are available nationally for use by amateur aircraft builders. Advertisements tend to be somewhat vague and may be misleading as to whether a kit is eligible for amateur-built certification. It is not advisable to order a kit prior to verifying with an FAA Inspector that the aircraft, upon completion, would be eligible for certification as amateur-built under existing rules and established policy.

(4) It should be noted that the FAA does not certify aircraft kits or approve kit manufacturing; however, the FAA does perform evaluations of kits that have potential for national sales interest, but only for the purpose of determining whether an aircraft built from the kit can meet the "major portion" criteria.

f. Safety Precaution Recommendations.

(1) All Aircraft.

(a) The pilot should thoroughly familiarize himself with the ground handling characteristics of the aircraft by conducting taxi tests before attempting flight operations.

(b) Before the first flight of an amateur-built aircraft, the pilot should take precautions to ensure that emergency equipment and personnel are readily available in the event of an accident.

(c) Violent (acrobatic) maneuvers should not be attempted until sufficient flight experience has been gained to establish that the aircraft is satisfactorily controllable throughout its normal ranges of speeds and maneuvers. Those maneuvers successfully demonstrated while in the test area may continue to be permitted by the FAA when the operating limitations are expanded.

(2) (Not Included.)

g. Documentation Requirements.

(1) To preclude any problems or questions concerning source or specification of materials, parts, etc., used in fabricating the aircraft, it would be helpful if the builder kept copies of all invoices or other shipping documents.

(2) A construction log maintained by the builder for the project, including photographs taken as major components are completed will be acceptable substantiation that the builder constructed the major portion of the aircraft.

(3) The aircraft's flight history should be recorded in an aircraft logbook. The nature as well as duration of each flight should be documented. If the aircraft is considered acrobatic, the acrobatic maneuvers should be demonstrated in the flight test area and recorded in the aircraft logbook.

8. RECURRENT AIRWORTHINESS CERTIFICATION. When an amateur-built aircraft has completed operations in the assigned test area, it is eligible for an UNLIMITED duration airworthiness certificate, an operating limitation airworthiness certificate. With the issuance of the unlimited airworthiness certificate, an operating limitation requiring a condition inspection at 12 month intervals is imposed. The aircraft builder can be certified as a repairman to enable him to perform the condition inspection. Specific information regarding the repairman certification can be found in AC No. 65-23, Certification of Repairman (Experimental Aircraft Builders).

9. REFERENCE MATERIAL.

a. Federal Aviation Regulations

Part 21—Certification Procedures for Products and Parts.

Part 45—Identification and Registration Marking.

Part 47—Aircraft Registration.

Part 65—Certification: Airmen Other than Flight Crew Members.

Part 91—General Operating and Flight Rules.

b. Advisory Circulars

20-90—Address List for Engineering and Manufacturing District Offices (Now redesignated Manufacturing Inspection District Offices.)

*43.13-1—Acceptable Methods, Techniques and Practices Aircraft Inspection and Repair.

*43.13-1 and 2A—Acceptable Methods, Techniques and Practices Aircraft Alterations.

65-23—Certification of Repairmen (Experimental Aircraft Builders).

*91-23—Pilot's Weight and Balance Handbook.

10. HOW TO GET PUBLICATIONS. The FAR and those ACs for which a fee is charges (marked with an asterisk (*) in paragraph 9(b)) may be obtained from the Superintendent of Documents, U.S. Government Printing Office, Washington, D.C. 20402. A listing of FARs and current prices is in AC 00-44, Status of Federal Aviation Regulations, and a listing of all ACs is in AC 00-2, Advisory Circular Checklist. These two ACs may be obtained free of charge from:

>Department of Transportation
>Publications Section, M-442.3
>Washington, D.C. 20590

Appendix 1 (AC 20-27C)

ADDRESSES

>Experimental Aircraft Association, Inc.
> Wittman Airfield
> Oshkosh, WI 54903
>Federal Aviation Administration
>Phone (405) 686-4331)

Mailing:	Airmen and Aircraft Registry Department of Transportation P.O. Box 25504 Oklahoma City, Oklahoma 73125
Street:	6500 South MacArthur Boulevard Oklahoma City, Oklahoma

PUBLICATIONS:

1. To request free advisory circulars:
>U.S. Department of Transportation
>Publications Section, M-442.32
>Washington, D.C. 20590

2. To be placed on FAA's mailing list for free advisory circulars:
>U.S. Department of Transportation
>Distribution Requirements Section, M-481.1
>Washington, D.C. 20590

3. To purchase for sale advisory circulars and FARs:

>Superintendent of Documents
>U.S. Government Printing Office
>Washington, D.C. 20402

The FAA allows amateur builders to repair and maintain their homebuilt airplanes. Circular 65-23, describing this policy is reproduced here:

CERTIFICATION OF REPAIRMEN (EXPERIMENTAL AIRCRAFT BUILDERS)

1. PURPOSE. This advisory circular (AC) provides guidance to builders of experimental aircraft concerning certification of repairmen.

2. RELATED FEDERAL AVIATION REGULATION (FAR). FAR Part 21, Sections 21.171, 21.173, 21.175, 21.177, 21.179, 21.181, 21.191, 21.193, and 21.195; FAR Part 43, Appendix D; FAR Part 65, Sections 65.1, 65.11, 65.12, 65.13, 65.15, 65.16, 65.20, 65.21, and 65.104; FAR Part 91, Section 91.42.

3. RELATED MATERIAL. (outdates)

4. BACKGROUND.

a. Previously, experimental aircraft certificates were effective for one year after the date of issuance or renewal, unless a shorter period was prescribed by the Administrator of the Federal Aviation Administration (FAA). Under the amended provisions of FAR Section 21.181(a) (3), effective September 10, 1979, experimental certificates issued to aircraft for the purpose of exhibiion, air racing, or operating amateur-built aircraft have an unlimited duration unless the Administrator finds that a specific period should be established. Thus, performance of recertification inspections on these aircraft by FAA inspectors will no longer be required. However, inspectors will continue to perform original certification inspections of experimental aircraft and construction inspections of amateur-built and amateur-built exhibition and air racing aircraft.

b. In conjunction with amended FAR Section

21.181(a) (3), a new FAR Section 65.104, Repairman Certificates (experimental aircraft builder)—Eligibility, Privileges, and Limitations, was added to FAR Part 65. This section provides that a qualified builder of each exhibition, air racing, and amateur-built aircraft may be certified as a repairman and would be privileged to perform condition inspections in accordance with Appendix D of FAR Part 43. However, aircraft manufacturing companies who produce experimental aircraft are not eligible for repairmen certificates.

c. When provided by the aircraft limitations, exhibition, air racing, and amateur-built aircraft may be inspected (condition inspections) by FAA certified mechanics holding airframe and powerplant rating or FAA certified and appropriately rated repair stations, in accordance with Appendix D of FAR Part 43.

5. ELIGIBILITY. An individual desiring to be certified as a repairman is required to:

a. Make application for a repairman certificate at the time of original certification. Builders who have had their aircraft certificated prior to the effective date (September 10, 1979) of revised FAR Sections 21.181(a) (3) and new 65.104 may make application for repairman certification prior to the next condition inspection due date.

b. Be a U.S. citizen or an individual of a foreign country who has been admitted for permanent residence in the United States.

c. Be 18 years of age, and the primary builder of the aircraft. For example, when a school, club, or a partnership builds an aircraft, only one individual will be considered for a repairman certificate for each aircraft built, such as the class instructor or designated project leader.

d. Demonstrate to the certificating FAA inspector his or her ability to perform condition inspections and to determine whether the subject aircraft is in a condition for safe operation.

Note: The eligibility requirements of FAR Section 65.104 are in no way associated with those eligibility requirements for repairman shown in FAR Section 65.101, titled "Eligibility Requirements: General."

6. PRIVILEGES AND LIMITATIONS. Holders of repairman certificates may perform "condition inspections" on specific aircraft built by the certificate holder. The aircraft will be identified by make, model, and serial number as shown on the repairman certificate. During the aircraft certification procedure, the FAA issues operating limitations, as required by FAR Section 91.42 of FAR Part 91, to ensure an adequate level of safety. These limitations require that the subject aircraft be inspected annually by a repairman, the holder of an FAA mechanic certificate with appropriate ratings (airframe and powerplant), or an appropriately rated FAA repair station. Condition inspections will be performed in accordance with the scope and detail of FAR Part 43, Appendix D.

Operating limitations will also require that an appropriate entry be made in the aircraft maintenance records to show performance of this limitation.

Note: It should be noted that the privileges and limitations of FAR Section 65.104 are in no way associated with those privileges and limitations of FAR Section 65.103, titled "Repairman Certificate: Privileges and Limitations."

7. APPLICATION.

a. Applicants may obtain copies of the Airman Certificate and/or Rating Application, FAA Form 8610-1 (OMB 04-R0065), from their local FAA General District Office or Flight Standards Office. Applicants must complete this form and submit it to their local FAA office. Appropriate Items I, III, and IV of this form must be completed. The box for "repairman" (at top of form) must be checked and underneath in the space for "specify rating" print or type the words "Experimental Aircraft Builder." Also, print or type in the "Type of Work Performed" of Item III the following information relating to the subject amateur-built aircraft:

Aircraft Make......................
Model............................
Serial Number.....................
Certification Date of Aircraft.........

Applicants should read the Privacy Act statement attached to FAA Form 8610-2 prior to completing this form.

b. When an applicant meets the certificate eligibility requirements, a Temporary Airman Certificate, FAA Form 8060-4, will be issued. Permanent certificates will be mailed to the holder of a Temporary Airman Certificate within 120 days of issuance.

8. SURRENDERED CERTIFICATE PROCEDURES. Repairman certificates should be surrendered whenever the aircraft is destroyed or sold. However, in the latter situation, the repairman may elect to retain the certificate in order to perform condition inspections on the aircraft for the new owner. Surrendered certificates should be forwarded to the Mike Monroney Aeronautical Center, Airmen Certification Branch, AAC-260, P.O. Box 25082, Oklahoma City, Oklahoma 73126, with a brief statement of reasons for surrender.

9. TYPICAL AIRCRAFT OPERATING LIMITATIONS. The following or similarly worded aircraft operating limitations may be issued at the time of aircraft certification:

a. No person may operate this aircraft unless within the preceding 12 calendar months it has had a condition inspection performed in accordance with Appendix D of FAR Part 43 and is found to be in a condition for safe operation. Additionally, this inspection shall be recorded in accordance with limitation d. listed below.

b. For amateur built and Amateur-built exhibition and air racing aircraft: Only FAA-certificated repairman (show repairman's name) mechanics holding an airframe and power plant rating, and appropriately rated repair stations may perform condition inspections in accordance with Appendix D of FAR Part 43.

c. For other exhibition and air racing aircraft: Only FAA-certified and rated airframe and powerplant mechanics and appropriately rated repair stations may perform condition inspections in accordance with Appendix D of FAR Part 43.

d. Condition inspections shall be recorded in the aircraft maintenance records showing the following or a similarly worded statement, "I certify that this aircraft has been inspected on (insert date) in accordance with the scope and detail of Appendix D of FAR Part 43 and found to be in a condition for safe operation." The entry will include the aircraft total time-in-service, the name, signature, and certificate type and number of the person performing the inspection.

DATE _____

AMATEUR-BUILT AIRCRAFT DATA

DOT/Federal Aviation Administration
Flight Standards District #62
P.O. Box 17325
Dulles International Airport
Washington, D.C. 20041

Please complete this form and mail it to this office

1. Aircraft Make or Type _____
2. Engine Make and Horsepower _____
3. Designer of Aircraft _____
4. Type of Construction _____
5. Materials Used (Aircraft or Otherwise) _____
6. Name of Builder _____
7. Address _____
8. Telephone No. (HOME) _____ (BUSINESS) _____
9. Address of Location of Project _____
10. Are You a Member of EAA?   YES ☐   NO ☐

    If "Yes", EAA Chapter Affiliation _____

11. Other Pertinent Information _____
    _____
    _____
    _____
    _____

NOTE: Please include a 3 view drawing of aircraft.

SIGNATURE _____

Please give driving instructions to project location on reverse side.

Fig. 9-1. A typical questionnaire used by a local FAA office during the initial builder contact. (courtesy FAA)

PRIVACY ACT OF 1974 (PL 93-579) requires that users of this form be informed of the authority which allows the solicitation of the information and whether disclosure of such information is mandatory or voluntary; the principal purpose for which the information is intended to be used; the routine uses which may be made of the information gathered; and the effects, if any, of not providing all or any part of the requested information.

The Federal Aviation Act of 1958 requires the registration of each United States civil aircraft as a prerequisite to its operation. An aircraft is eligible for registration only: (1) if it is not registered under the laws of any foreign country; and (2) if it is owned by (a) a citizen of the United States; or (b) an individual citizen of a foreign country who has lawfully been admitted for permanent residence in the United States; or (c) a corporation lawfully organized and doing business under the laws of the United States or any State thereof so long as such aircraft is based and primarily used in the United States; or (d) a governmental unit. Operation of an aircraft that is not registered may subject the operator to a civil penalty.

This form identifies the aircraft to be registered, and provides the name and permanent address for mailing the registration certificate. Incomplete submission will prevent or delay issuance of your registration certificate.

The following routine uses are made of the information gathered:

(1) To determine that aircraft are registered in accordance with the provisions of the Federal Aviation Act of 1958.

(2) To support investigative efforts of investigation and law enforcement agencies of Federal, State and foreign governments.

(3) To serve as a repository of legal documents used by individuals and title search companies to determine the legal ownership of an aircraft.

(4) To provide aircraft owners and operators information about potential mechanical defects or unsafe conditions of their aircraft in the form of airworthiness directives.

(5) To provide supporting information in court cases concerning liability of individuals in law suits.

(6) To serve as a data source for management information for production of summary descriptive statistics and analytical studies in support of agency functions for which the records are collected and maintained.

(7) To respond to general requests from the aviation community or the public for statistical information under the Freedom of Information Act or to locate specific individuals or specific aircraft for accident investigation, violation, or other safety related requirements.

(8) To provide data for the automated aircraft registration master file.

(9) To provide documents for microfiche backup record.

(10) To provide data for development of the aircraft registration statistical system.

(11) To operate an aircraft register in magnetic tape and publication form required by the International Civil Aviation Organization (ICAO) agreement containing information on aircraft by registration number, type of aircraft, and name and address of owners used for internal FAA safety program purposes.

(12) The aircraft records maintained by the FAA Aircraft Registry are public records and are open for inspection in Room 123 of the Aviation Records Building, Mike Monroney Aeronautical Center, 6500 South MacArthur, Oklahoma City, Oklahoma 73125. Individuals interested in such information may make a personal search of the records or may avail themselves of the services of a company or an attorney.

Fig. 9-2. Privacy Act of 1974 statement about aircraft registration. (courtesy FAA)

FORM APPROVED
OMB NO. 2120-0029
EXP. DATE 10/31/84

UNITED STATES OF AMERICA DEPARTMENT OF TRANSPORTATION
FEDERAL AVIATION ADMINISTRATION-MIKE MONRONEY AERONAUTICAL CENTER
AIRCRAFT REGISTRATION APPLICATION

CERT. ISSUE DATE

UNITED STATES REGISTRATION NUMBER **N**

AIRCRAFT MANUFACTURER & MODEL

AIRCRAFT SERIAL No.

FOR FAA USE ONLY

TYPE OF REGISTRATION (Check one box)

☐ 1. Individual ☐ 2. Partnership ☐ 3. Corporation ☐ 4. Co-owner ☐ 5. Gov t. ☐ 8. Foreign-owned Corporation

NAME OF APPLICANT (Person(s) shown on evidence of ownership. If individual, give last name, first name, and middle initial.)

TELEPHONE NUMBER: (   ) —
ADDRESS (Permanent mailing address for first applicant listed.)

Number and street: _____

Rural Route: _____   P.O. Box: _____

CITY	STATE	ZIP CODE

☐ **CHECK HERE IF YOU ARE ONLY REPORTING A CHANGE OF ADDRESS**
ATTENTION! Read the following statement before signing this application.

A false or dishonest answer to any question in this application may be grounds for punishment by fine and / or imprisonment (U.S. Code, Title 18, Sec. 1001).

CERTIFICATION

I/WE CERTIFY:

(1) That the above aircraft is owned by the undersigned applicant, who is a citizen (including corporations) of the United States.

(For voting trust, give name of trustee: _____), or:

CHECK ONE AS APPROPRIATE:

a. ☐ A resident alien, with alien registration (Form 1-151 or Form 1-551) No. _____

b. ☐ A foreign-owned corporation organized and doing business under the laws of (state or possession) _____, and said aircraft is based and primarily used in the United States. Records of flight hours are available for inspection at _____

(2) That the aircraft is not registered under the laws of any foreign country; and
(3) That legal evidence of ownership is attached or has been filed with the Federal Aviation Administration.

NOTE: If executed for co-ownership all applicants must sign. Use reverse side if necessary.

TYPE OR PRINT NAME BELOW SIGNATURE

EACH PART OF THIS APPLICATION MUST BE SIGNED IN INK.

SIGNATURE	TITLE	DATE
SIGNATURE	TITLE	DATE
SIGNATURE	TITLE	DATE

NOTE: Pending receipt of the Certificate of Aircraft Registration, the aircraft may be operated for a period not in excess of 90 days, during which time the PINK copy of this application must be carried in the aircraft.

AC FORM 8050-1 (1-83) (0052-00-628-9005)

Fig. 9-3. AC Form 8050-1 Aircraft Registration Application. (courtesy FAA)

Fig. 9-4. AC Form 8050-2 Aircraft Bill of Sale. (courtesy FAA)

```
            AFFIDAVIT OF OWNERSHIP FOR AMATEUR-BUILT AIRCRAFT

U.S. Identification Number  N2345X

Builder's Name  Charles E. Griffin

Model   CFG-1              Serial Number   00001

Class (airplane, rotorcraft, glider, etc.)  airplane

Type of Engine Installed (reciprocating, turbopropeller, etc.)
        turbopropeller

Number of Engines Installed    1

Manufacturer, Model, and Serial Number of each Engine Installed
        Twister, PHP, 5064


Built for Land or Water Operation   land

Number of Seats    1

The above-described aircraft was built from parts by the undersigned and
I am the owner.


                                    Charles E. Griffin
                                    (Signature of Owner-Builder)

State of  Oklahoma

County of  Oklahoma

Subscribed and sworn to before me this  23  day of  September  , 19 80 .

My commission expires    12/16/82          .


A. B. Jackson
(Signature of Notary Public)
```

Fig. 9-5. Affidavit of Ownership for Amateur-Built Aircraft. (courtesy FAA)

DEPARTMENT OF TRANSPORTATION
FEDERAL AVIATION ADMINISTRATION

DATE: September 30, 1980

IN REPLY REFER TO: AAC-250

AERONAUTICAL CENTER
P O BOX 25504
OKLAHOMA CITY OKLAHOMA 73125

SUBJECT: Identification number assignment and registration of amateur-built aircraft; your ltr of 9/25/80

FROM: FAA Aircraft Registry

TO: Beverly J. Johnson
President, Johnson Safety Corporation
Post Office Box 61479
Oklahoma City, Oklahoma 73100

Dear Mrs Johnson:

[xx] U.S. identification number N <u>2345X</u> has been ASSIGNED.

[ ] A U.S. identification number of <u>our</u> choice may be assigned, free of charge, to your amateur-built aircraft when you submit a complete description of the aircraft. The form on the back of this letter may be used, as it meets our requirements for both description and registration purposes. Authority to use a number assigned free of charge expires 90 days after the date it is issued unless the aircraft is registered within that period.

[ ] U.S. identification number N_____ has been RESERVED.

[ ] A U.S. identification number of <u>your</u> choice may be reserved, if available, for one year by sending a written request and a $10 fee for each number to be reserved. Please list five numbers, in order of preference, in case your first choice is not available. If the number is not assigned to an aircraft prior to the end of the year, the reservation will expire, but may be renewed from year to year upon request and payment of a $10 renewal fee.
    NOTICE: <u>This number may not be assigned or painted on an aircraft until approval is received from this office.</u>
Your written request to assign the reserved number to a particular aircraft must include a complete description of the aircraft. The form on the back of this letter may be used.

[ ] The items checked below are required to complete registration of your amateur-built aircraft:

    [ ] Completed and signed Aircraft Registration Application; for is enclosed.

    [ ] Registration of ownership, signed before a notary public, and showing a description of the aircraft. The form on the back of this letter meets
    [ ] our requirements and may be used if you wish. If the aircraft is built from a kit, a bill of sale from the kit manufacturer is also needed.

    See items checked on the reverse side. These must be included on the enclosed affidavit.
    [ ]

Records Examiner
AC Form 8050-88 (9-75) (0052-00-559-0002) Supersedes previous edition

Fig. 9-6. AC Form 8050-88 Identification Number Assignment. (courtesy FAA)

Form Approved
O.M.B. No. 2120-0018

US Department of Transportation
Federal Aviation Administration

APPLICATION FOR AIRWORTHINESS CERTIFICATE

INSTRUCTIONS — Print or type. Do not write in shaded areas; these are for FAA use only. Submit original only to an authorized FAA Representative. If additional space is required, use an attachment. For special flight permits complete Sections II and VI or VII as applicable.

I. AIRCRAFT DESCRIPTION

1. REGISTRATION MARK	2. AIRCRAFT BUILDER'S NAME *(Make)*	3. AIRCRAFT MODEL DESIGNATION	**FAA CODING**
4. AIRCRAFT SERIAL NO.	5. ENGINE BUILDER'S NAME *(Make)*	6. ENGINE MODEL DESIGNATION	
7. NUMBER OF ENGINES	8. PROPELLER BUILDER'S NAME *(Make)*	9. PROPELLER MODEL DESIGNATION	10. AIRCRAFT IS *(Check if applicable)* IMPORT

II. CERTIFICATION REQUESTED

APPLICATION IS HEREBY MADE FOR: *(Check applicable items)*

A 1 STANDARD AIRWORTHINESS CERTIFICATE *(Indicate category)* NORMAL | UTILITY | ACROBATIC | TRANSPORT | GLIDER | BALLOON

B SPECIAL AIRWORTHINESS CERTIFICATE *(Check appropriate items)*

2	LIMITED						
5	PROVISIONAL *(Indicate class)*	1	CLASS I				
		2	CLASS II				
3	RESTRICTED *(Indicate operation(s) to be conducted)*	1	AGRICULTURE AND PEST CONTROL	2	AERIAL SURVEYING	3	AERIAL ADVERTISING
		4	FOREST *(Wildlife conservation)*	5	PATROLLING	6	WEATHER CONTROL
		7	CARRIAGE OF CARGO	0	OTHER *(Specify)*		
4	EXPERIMENTAL *(Indicate operation(s) to be conducted)*	1	RESEARCH AND DEVELOPMENT	2	AMATEUR BUILT	3	EXHIBITION
		4	RACING	5	CREW TRAINING		MKT. SURVEY
		0	TO SHOW COMPLIANCE WITH FAR				
8	SPECIAL FLIGHT PERMIT *(Indicate operation to be conducted, then complete Section VI or VII as applicable on reverse side)*	1	FERRY FLIGHT FOR REPAIRS, ALTERATIONS, MAINTENANCE OR STORAGE				
		2	EVACUATE FROM AREA OF IMPENDING DANGER				
		3	OPERATION IN EXCESS OF MAXIMUM CERTIFICATED TAKE-OFF WEIGHT				
		4	DELIVERING OR EXPORT	5	PRODUCTION FLIGHT TESTING		
		6	CUSTOMER DEMONSTRATION FLIGHTS				

C 6 MULTIPLE AIRWORTHINESS CERTIFICATE *(Check ABOVE "Restricted Operation" and "Standard" or "Limited" as applicable.)*

III. OWNER'S CERTIFICATION

A. REGISTERED OWNER *(As shown on certificate of aircraft registration)* — **IF DEALER, CHECK HERE →**

NAME	ADDRESS

B. AIRCRAFT CERTIFICATION BASIS *(Check applicable blocks and complete items as indicated)*

AIRCRAFT SPECIFICATION OR TYPE CERTIFICATE DATA SHEET *(Give No. and Revision No.)*	AIRWORTHINESS DIRECTIVES *(Check if all applicable AD's complied with and give latest AD No.)*
AIRCRAFT LISTING *(Give page number(s))*	SUPPLEMENTAL TYPE CERTIFICATE *(List number of each STC incorporated)*

C. AIRCRAFT OPERATION AND MAINTENANCE RECORDS

CHECK IF RECORDS IN COMPLIANCE WITH FAR 91.173	TOTAL AIRFRAME HOURS	3 EXPERIMENTAL ONLY *(Enter hours flown since last certificate issued or renewed)*

D. **CERTIFICATION** — I hereby certify that I am the registered owner (or his agent) of the aircraft described above, that the aircraft is registered with the Federal Aviation Administration in accordance with Section 501 of the Federal Aviation Act of 1958, and applicable Federal Aviation Regulations, and that the aircraft has been inspected and is airworthy and eligible for the airworthiness certificate requested.

DATE OF APPLICATION	NAME AND TITLE *(Print or type)*	SIGNATURE

IV. INSPECTION AGENCY VERIFICATION

A. THE AIRCRAFT DESCRIBED ABOVE HAS BEEN INSPECTED AND FOUND AIRWORTHY BY *(Complete this section only if FAR 21.183(d) applies)*

2 FAR PART 121 OR 127 CERTIFICATE HOLDER *(Give Certificate No.)*	3 CERTIFICATED MECHANIC *(Give Certificate No.)*	6 CERTIFICATED REPAIR STATION *(Give Certificate No.)*
5 AIRCRAFT MANUFACTURER *(Give name of firm)*		
DATE	TITLE	SIGNATURE

V. FAA REPRESENTATIVE CERTIFICATION

(Check ALL applicable blocks in items A and B)

A. I find that the aircraft described in Section I or VII meets requirements for — THE CERTIFICATE REQUESTED; 4 AMENDMENT OR MODIFICATION OF CURRENT AIRWORTHINESS CERTIFICATE

B. Inspection for a special flight permit under Section VII was conducted by: FAA INSPECTOR; FAA DESIGNEE; CERTIFICATE HOLDER UNDER — FAR 65 | FAR 121, 127 or 135 | FAR 145

DATE	DISTRICT OFFICE	4 DESIGNEE'S SIGNATURE AND NO.	1 FAA INSPECTOR'S SIGNATURE

FAA Form 8130-6 (12-83) SUPERSEDES PREVIOUS EDITION

Fig. 9-7. FAA Form 8130-6 Application for Airworthiness Certificate (front). (courtesy FAA)

VI. PRODUCTION FLIGHT TESTING

A. MANUFACTURER

NAME	ADDRESS

B. PRODUCTION BASIS *(Check applicable item)*

PRODUCTION CERTIFICATE *(Give production certificate number)*	
TYPE CERTIFICATE ONLY	
APPROVED PRODUCTION INSPECTION SYSTEM	

C. GIVE QUANTITY OF CERTIFICATES REQUIRED FOR OPERATING NEEDS ⟶

DATE OF APPLICATION	NAME AND TITLE *(Print or type)*	SIGNATURE

VII. SPECIAL FLIGHT PERMIT PURPOSES OTHER THAN PRODUCTION FLIGHT TEST

A. DESCRIPTION OF AIRCRAFT

REGISTERED OWNER	ADDRESS
BUILDER *(Make)*	MODEL
SERIAL NUMBER	REGISTRATION MARK

B. DESCRIPTION OF FLIGHT   CUSTOMER DEMONSTRATION FLIGHTS ☐ *(Check if applicable)*

FROM	TO
VIA	DEPARTURE DATE	DURATION

C. CREW REQUIRED TO OPERATE THE AIRCRAFT AND ITS EQUIPMENT.

PILOT	CO-PILOT	NAVIGATOR	OTHER *(Specify)*

D. THE AIRCRAFT DOES NOT MEET THE APPLICABLE AIRWORTHINESS REQUIREMENTS AS FOLLOWS:

E. THE FOLLOWING RESTRICTIONS ARE CONSIDERED NECESSARY FOR SAFE OPERATION *(Use attachment if necessary)*.

F. CERTIFICATION — I hereby certify that I am the registered owner (or his agent) of the aircraft described above, that the aircraft is registered with the Federal Aviation Administration in accordance with Section 501 of the Federal Aviation Act of 1958, and applicable Federal Aviation Regulations, and that the aircraft has been inspected and is airworthy for the flight described.

DATE	NAME AND TITLE *(Print or type)*	SIGNATURE

VIII. AIRWORTHINESS DOCUMENTATION (FAA use only)

A. Operating Limitations and Markings in Compliance with FAR 91.31 as Applicable	**G.** Statement of Conformity, FAA Form 8130-9 *(Attach when required)*
B. Current Operating Limitations Attached	**H.** Foreign Airworthiness Certification for Import Aircraft *(Attach when required)*
C. Data, Drawings, Photographs, etc. *(Attach when required)*	**I.** Previous Airworthiness Certificate Issued in Accordance with
D. Current Weight and Balance Information Available in Aircraft	FAR _____ CAR _____ *(Original Attached)*
E. Major Repair and Alteration, FAA Form 337 *(Attach when required)*	**J.** Current Airworthiness Certificate Issued in Accordance with
F. This Inspection Recorded in Aircraft Records	FAR _____ *(Copy attached)*

Fig. 9-8. FAA Form 8130-6 (back). (courtesy FAA)

SAMPLE ONLY

UNITED STATES OF AMERICA

DEPARTMENT OF TRANSPORTATION - FEDERAL AVIATION ADMINISTRATION

SPECIAL AIRWORTHINESS CERTIFICATE

A	CATEGORY/DESIGNATION		
	PURPOSE		
B	MANU-FACTURER	NAME	
		ADDRESS	
C	FLIGHT	FROM	
		TO	
D	N—		SERIAL NO.
	BUILDER		MODEL
E	DATE OF ISSUANCE		EXPIRY
	OPERATING LIMITATIONS DATED		ARE A PART OF THIS CERTIFICATE
	SIGNATURE OF FAA REPRESENTATIVE		DESIGNATION OR OFFICE NO.

Any alteration, reproduction or misuse of this certificate may be punishable by a fine not exceeding $1,000 or imprisonment not exceeding 3 years, or both. THIS CERTIFICATE MUST BE DISPLAYED IN THE AIRCRAFT IN ACCORDANCE WITH APPLICABLE FEDERAL AVIATION REGULATIONS.

FAA FORM 8130-7 (10/82) *SEE REVERSE SIDE*

Fig. 9-9. FAA Form 8130 Special Airworthiness Certificate. (courtesy FAA)

Federal Communications Commission
Gettysburg, PA 17325

Approved by OMB
3060-0040
Expires 3/31/86

APPLICATION FOR AIRCRAFT RADIO STATION LICENSE

- Read Instructions above before completing application.
- Use typewriter or print clearly in ink.
- Sign and date application.
- Place First Class Postage on the reverse side of the card and mail.

1. FAA Registration or FCC Control Number. (If FAA Registration is not required for your aircraft, explain in Item 8.) N

2. Is application for a fleet license? ☐ No ☐ Yes
If yes give the number of aircraft in fleet, including planned expansion:

3. Type of applicant (check one)
☐ I-Individual
☐ P-Partnership
☐ A-Association
☐ C-Corporation
☐ D-Individual with Business Name
☐ G-Governmental Entity

4. Applicant/Licensee Name (See Instructions)

5. Mailing Address (Number and Street, P.O. Box or Route No., City, State, ZIP Code)

6. Frequencies Requested (check appropriate box(es) in 6.A. and/or 6.B.)

6A. DO NOT CHECK BOTH BOXES
☐ A-Private Aircraft ☐ C-Air Carrier

6.B. ADDITIONAL INFORMATION IS REQUIRED IF YOU CHECK HERE (See Instructions)
☐ T-Flight Test HF ☐ V-Flight Test VHF ☐ O-Other (Specify)

7. Application is for:
☐ New Station ☐ Renewal
☐ Modification

8. Answer space for any required statements

9. READ CAREFULLY BEFORE SIGNING: 1. Applicant waives any claim to the use of any particular frequency regardless of prior use by license or otherwise. 2. Applicant will have unlimited access to the radio equipment and will control access to exclude unauthorized persons. 3. Neither applicant nor any member thereof is a foreign government or representative thereof. 4. Applicant certifies that all statements made in this application and attachments are true, complete, correct and made in good faith. 5. Applicant certifies that the signature is that of the individual, or partner, or officer or duly authorized employee of a corporation, or officer who is a member of an unincorporated association, or appropriate elected or appointed official on behalf of a governmental entity.

WILLFUL FALSE STATEMENTS MADE ON THIS FORM ARE PUNISHABLE BY FINE AND/OR IMPRISONMENT U.S. CODE TITLE 18, SECTION 1001.

10. Signature

Date

FCC 404
June 1983

Fig. 9-10. FCC Form 404 Application for Aircraft Radio Station License. (courtesy FCC)

| Federal Communications Commission
Gettysburg, PA 17325 | TEMPORARY AIRCRAFT RADIO STATION
OPERATING AUTHORITY | Approved by OMB
3060-0040
Expires 3/31/86 |

Use this form if you want a temporary operating authority while your regular application, FCC Form 404, is being processed by the FCC. This authority authorizes the use of transmitters operating on the appropriate frequencies listed in Part 87 of the Commission's Rules.

- DO NOT use this form if you already have a valid aircraft station license.
- DO NOT use this form when renewing your aircraft license.
- DO NOT use this form if you are applying for a fleet license.
- DO NOT use this form if you do not have an FAA Registration Number.

ALL APPLICANTS MUST CERTIFY:

1. I am not a representative of a foreign government.
2. I have applied for an Aircraft Radio Station License by mailing a completed FCC Form 404 to the Federal Communication's Commission, P.O. Box 1030, Gettysburg, PA 17325.
3. I have not been denied a license or had my license revoked by the FCC.
4. I am not the subject of any adverse legal action concerning the operation of a radio station.
5. I will ensure that the Aircraft Radio Station will be operated by an individual holding the proper class of license or permit required by the Commission's Rules.

WILLFUL FALSE STATEMENTS VOID THIS PERMIT AND ARE PUNISHABLE BY FINE AND/OR IMPRISONMENT.

Name of Applicant (Print or Type)	Signature of Applicant
FAA Registration Number (Use as Temporary Call Sign)	Date Form 404 Mailed to FCC

Your authority to operate your Aircraft Radio Station is subject to all applicable laws, treaties and regulations and is subject to the right of control of the Government of the United States. This authority is valid for 90 days from the date the Form 404 is mailed to the FCC.

YOU MUST POST THIS TEMPORARY AUTHORITY ON BOARD YOUR AIRCRAFT

DETACH HERE — DO NOT MAIL THIS PART

FCC 404-A
June 1983

Fig. 9-11. FCC Form 404, detachable portion giving temporary operating authority. (courtesy FCC)

No certificate may be issued unless a completed application form has been received (14 C.F.R. 65).

Form Approved OMB No. 04-R0065

DEPARTMENT OF TRANSPORTATION – FEDERAL AVIATION ADMINISTRATION

AIRMAN CERTIFICATE AND/OR RATING APPLICATION

☐ MECHANIC ☐ AIRFRAME ☐ POWERPLANT

☐ REPAIRMAN ______ *(Specify Rating)*

☐ PARACHUTE RIGGER ☐ SENIOR ☐ MASTER ☐ SEAT ☐ CHEST ☐ BACK ☐ LAP

APPLICATION FOR ☐ ORIGINAL ISSUANCE ☐ ADDED RATING

I. APPLICANT INFORMATION

A. NAME *(First, Middle, Last)*

B. SOCIAL SECURITY NO. | C. DOB *(Mo., Day, Yr.)* | D. HEIGHT IN | E. WEIGHT

F. HAIR | G. EYES | H. SEX | I. NATIONALITY

J. PLACE OF BIRTH

K. PERMANENT MAILING ADDRESS NUMBER AND STREET, P.O. BOX, ETC

CITY, STATE, ZIP CODE

COUNTY

L. HAVE YOU EVER HAD AN AIRMAN CERTIFICATE DENIED, SUSPENDED, OR REVOKED? ☐ NO ☐ YES → If "Yes," explain on an attached sheet keying to appropriate item number

M. DO YOU NOW OR HAVE YOU EVER HELD AN FAA AIRMAN CERTIFICATE? ☐ NO ☐ YES SPECIFY TYPE

N. HAVE YOU EVER BEEN CONVICTED FOR VIOLATION OF ANY FEDERAL OR STATE STATUTES PERTAINING TO NARCOTIC DRUGS, MARIHUANA, AND DEPRESSANT OR STIMULANT DRUGS OR SUBSTANCES? ☐ NO ☐ YES → DATE OF FINAL CONVICTION

II. CERTIFICATE OR RATING APPLIED FOR ON BASIS OF –

☐ A. CIVIL EXPERIENCE ☐ B. MILITARY EXPERIENCE ☐ C. LETTER OF RECOMMENDATION FOR REPAIRMAN *(Attach copy)*

☐ D. GRADUATE OF APPROVED COURSE: (1) NAME AND LOCATION OF SCHOOL | (2) SCHOOL NO. | (3) CURRICULUM FROM WHICH GRADUATED | (4) DATE

☐ E. STUDENT HAS MADE SATISFACTORY PROGRESS AND IS RECOMMENDED TO TAKE THE ORAL-PRACTICAL TEST (FAR 65.80): (1) SCHOOL NAME | NO. | (2) SCHOOL OFFICIAL'S SIGNATURE

☐ F. SPECIAL AUTHORIZATION TO TAKE MECHANIC'S ORAL PRACTICAL TEST (FAR 65.80): (1) DATE AUTH | (2) DATE AUTH EXPIRES | (3) FAA INSPECTOR SIGNATURE | (4) FAA DIST OFC

III. RECORD OF EXPERIENCE

A. MILITARY COMPETENCE OBTAINED IN → (1) SERVICE | (2) RANK OR PAY LEVEL | (3) MILITARY SPECIALTY CODE

B. APPLICANTS OTHER THAN FAA CERTIFICATED SCHOOL GRADUATES, LIST EXPERIENCE RELATING TO CERTIFICATE AND RATING APPLIED FOR *(Continue on separate sheet if more space is needed)*

DATES–MONTH AND YEAR FROM	TO	EMPLOYER AND LOCATION	TYPE WORK PERFORMED

C. PARACHUTE RIGGER APPLICANTS: INDICATE BY TYPE HOW MANY PARACHUTES PACKED → SEAT | CHEST | BACK | LAP

FOR MASTER RATING ONLY — PACKED AS A – ☐ SENIOR RIGGER ☐ MILITARY RIGGER

IV. APPLICANT'S CERTIFICATION

I CERTIFY THAT THE STATEMENTS BY ME ON THIS APPLICATION ARE TRUE

A. SIGNATURE | B. DATE

FOR FAA USE ONLY

Emp | reg | D.O. | seal | con | iss | Act | lev | TR | s.h. | Srch | # rte | RATING (1) | RATING (2) | RATING (3) | RATING (4)

LIMITATIONS

FAA Form 8610-2 (10-76) SUPERSEDES FAA FORM 8310-2 AND FAA FORM 8000-33

Fig. 9-12. FAA Form 8610-2 Airman Certificate and/or Rating Application. (courtesy FAA)

Chapter 10

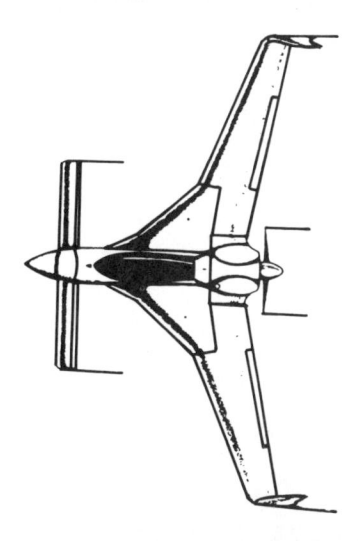

EAA Assistance

The EAA (Experimental Aircraft Association) is an organization, over 30 years old, dedicated to the aviation enthusiast, the individuals who fly for sport, pleasure, and education. The EAA is known for being a "can-do" organization.

AREAS OF INTEREST

The EAA and its members can be found involved in any or all of the following areas of sport flying:

- ☐ Amateur building.
- ☐ Restorations.
- ☐ Ultralights.
- ☐ Antiques.
- ☐ Warbirds.
- ☐ Classics.
- ☐ Vintage.
- ☐ Aerobatics.
- ☐ Rotorcraft.

INFORMATION SERVICES

The EAA's monthly magazine, *Sport Aviation*, is my personal favorite of all the current aviation magazines available today. It is devoted to the individual who loves airplanes, flies airplanes, builds and rebuilds airplanes, or is just plain interested in airplanes!

Sport Aviation covers everything about flying on the non-commercial level, from safety through construction advice. Also included are reviews of recent and old airplanes, as well as big news stories such as the annual Oshkosh fly-in.

In addition to the *Sport Aviation* magazine, the EAA also publishes magazines for special interest groups (i.e., the *Sport Aerobatics, Vintage Airplane, Ultralight,* and *Warbirds*), a complete series of technical "how-to-do-it" manuals, and various slide/movies/video sets, the latter covering gatherings and events such as fly-ins.

GATHERINGS

The Oshkosh Fly-In is held in the summer at EAA headquarters in Wisconsin. The official name for Oshkosh is the Annual EAA International Fly-In Convention and Sport Aviation Exhibition, and is co-sponsored by the EAA and the EAA Aviation Foundation. It is the "Mecca" of sport flying. If you have never been to Oshkosh, *go* (Figs. 10-1 through 10-3).

Other airshows, or fly-ins, are sponsored by local EAA Chapters throughout the country. Although much smaller than Oshkosh, they are certainly well worth attending. Most have business meetings and forums presented by local members or representatives from the FAA.

LOCAL CHAPTERS

The EAA has better than 700 chapters all over the world. Most locales in the United States will have a chapter. For further information about chapters, and how to find them, contact the EAA and ask for the EAA International Chapter Directory.

DIVISIONS

The EAA has four divisions:

- ☐ Antique/Classic Division, involved with the restoration and preservation of the planes of yesteryear.
- ☐ International Aerobatic Club, which sponsers aerobatic competitions.
- ☐ Warbirds of America, whose rally cry is "Keep 'Em Flying," with regards to the famous military fighters and bombers of WWII.

Fig. 10-1. One word describes this scene: Oshkosh! (photo by Jim Koepnick, courtesy EAA)

Fig. 10-2. There is something for everyone at Oshkosh. (photo by Ted Koston, courtesy EAA)

- EAA Ultralight Association, dealing with ultralights only.

ASSISTANCE

Of course, most readers' main concern is the EAA's involvement with homebuilts. This is where they really shine. Each local chapter is made up of individuals who are building, have built, or want to build or rebuild airplanes. Among the members of these chapters you will find loads of experience, much advice, assistance when you need it, and the friendship of others with interests similar to yours.

As part of the EAA support to the homebuilder there is the EAA Designee Program. There are nearly 1000 Designees nationwide whose "job" it is to assist the homebuilder in completing a safe airplane.

The Designee(s) are members of local EAA Chapters who have many years of experience in building/rebuilding airplanes. Often they have FAA mechanic or inspector ratings (A&E or AI). These Designees volunteer their services to EAA members.

Although the Designee has no power with the FAA, he is recognized for his expertise and experience. The FAA strongly urges those building an aircraft to utilize their services.

HISTORY OF THE EAA

The following is a complete chronology of the EAA, and is reprinted here by permission.

1953, January

Paul Howard Poberezny, the leader of a small group of aviation enthusiasts who had been meeting at his home, founded the Experimental Aircraft Association and is elected its first and only President. On January 26, Poberezny calls the first official meeting of the EAA at Curtis-Wright Field. The membership immediately expands from six to

Fig. 10-3. Almost heaven—airplanes as far as the eye can see belong to visitors to Oshkosh. (courtesy EAA)

31. The group originally gathers together to aid and assist amateur aircraft builders. However, its purposes quickly encompass the promotion of all facets of aviation—especially sport aviation—the preservation of America's rich heritage of personal flight, and the promotion of aviation safety.

The organization derives its name from the "Experimental Aircraft" category which is assigned to airplanes which are used for recreational and educational purposes only. One of the keys to the association's rapid and constant growth is the fact that its membership is open to anyone interested in aviation.

1953, February

The first issue of the official EAA newsletter is published. The *Experimenter* is originally written, typed, and mimeographed in Paul and Audrey Poberezny's basement. The *Experimenter* is destined to evolve into *Sport Aviation*, the most authoritative and widely respected aviation publication of its kind.

1953, September

The First Annual EAA Fly-In Convention had a two-fold purpose. It was, of course, the official business and social gathering of the fledgling EAA. It was also an effort by Paul Poberezny to bolster an ailing Milwaukee "air pageant." Although the first EAA fly-in was considered an unqualified success at the time, it was a humble affair when compared to today's EAA Convention, which is the aviation world's premier event.

1953, October

EAA member Ray Stits requests permission to establish an EAA Chapter in Riverside, California. EAA Chapter #1 is the first of over 700 worldwide Chapters which will eventually provide local activities for aviation enthusiasts and retain the "grass roots" ambiance of the sophisticated international organization. The chapter network is also largely responsible for the continuation of EAA's legendary volunteerism and camaraderie.

1954

Feature articles on the EAA begin appearing in *Flying* magazine and *Mechanix Illustrated*. These early feature articles are responsible for drawing attention to the new organization and, in turn, boosting its membership significantly. These early articles presage the many volumes of publicity later generated by and for the EAA and its activities.

1955, May, June, July

Mechanix Illustrated magazine features a series of articles on the construction of the Baby Ace. Paul Poberezny had modified and improved the design of the original Corben Baby Ace when he obtained control of the defunct Corben Airplane Company. The Baby Ace was now a suitable project for amateur homebuilders, and the articles on the airplane and its construction generated intense interest in the EAA, Paul Poberezny, and the Baby Ace by the legions of "do-it-yourselfers" in America. The result, of course, was a dramatic increase in EAA membership.

1955

EAA's Project Schoolflight is born when St. Rita's School in Chicago, Illinois, uses Poberezny's updated Baby Ace as an aviation project. Eventually, Project Schoolflight will be taken over by the nonprofit EAA Aviation Foundation and become the fastest growing aviation education program in the world. As a result of this outreach program, there are nearly 500 amateur-built aircraft under construction in various schools with approximately 175 airplanes already completed. The program is growing at a rate of about 70 new planes a year. Literally thousands of one-time Project Schoolflight students are now involved in the aviation industry. Thousands of others have a deeper and more sympathetic understanding of aviation.

1958, January

The first issue of *Sport Aviation*, EAA's flagship publication, is mailed to EAA members. *Sport Aviation* is the direct descendant of the *Experimenter*.

1959, February

The current untralight movement is predicted

in *Sport Aviation* articles by S.D. Coleman and Joe Kirk.

1959, August

The Annual EAA Fly-In Convention, although it is only seven years old, outgrows its convention site at Curtis-Wright Field in Milwaukee. Rockford, Illinois has been selected as the new site of the annual fly-in convention which is now attracting hundreds of planes and thousands of aviation enthusiasts.

1962, April

The EAA Air Museum Foundation is incorporated to permit tax-deductible donations. The new EAA Air Museum Foundation will eventually develop the world's largest private collection of aircraft and aviation-related artifacts. In addition to the new Museum, the EAA Air Museum Foundation will be responsible for EAA's outreach programs, research and development, educational programs, and it will be the cosponser of the annual EAA Fly-In Convention.

1964, August

EAA Headquarters is moved from the basement of Paul and Audrey Poberezny's home to a new building on property acquired in Franklin, Wisconsin, by the EAA.

1966, December

The EAA offices have quickly outgrown their new headquarters building. As a result, a new museum, office complex and restoration facility has been added to the original building and the EAA Headquarters site.

1970, August

The annual EAA Fly-In Convention has grown to include a total sport aviation exhibition. The event is being regularly attended by nearly 300,000 people and almost 10,000 airplanes. The convention site in Rockford, Illinois, can no longer accommodate the annual celebration of flight. Oshkosh, Wisconsin, is selected as the new permanent site for the world's largest aviation event.

1971

The EAA Aviation Foundation embarks on a research and development program directed at proving the utility, economy, and safety of using unleaded automobile fuel in certain types of aircraft.

1971, August

Because many EAA members are interested in very specific types of aviation activities, three EAA divisions are established to cater to their specialized activities. They include the Antique/Classic Division, the International Aerobatic Club, and the Warbirds of America. Each of these divisions has its own board of directors and newsletter. The newsletters will eventually develop into full-fledged magazines.

1976, August

The Annual EAA Fly-In Convention and Sport Aviation Exhibition is universally proclaimed the world's largest and greatest aviation event. It has also developed into the world's largest annual convention. The 1976 Convention is a milestone in the history of aviation because John Moody, the "father" of the modern ultralight movement, displays his powered hang glider for the first time before a large audience.

1977

On the 50th anniversary of Charles Lindbergh's historic solo Atlantic flight, an exceedingly accurate reproduction of the *Spirit of St. Louis*, built by the EAA Aviation Foundation's restoration shop, embarks on a cross-country commemorative tour. This tour was aimed at rekindling America's interest in its aviation history.

1978

The EAA Aviation Foundation, through Project Schoolflight, commissions an exact replica of Orville and Wilbur Wright's Flyer to be constructed by Blackhawk Technical School in Wisconsin. The Flyer will eventually become the centerpiece of the EAA Aviation Foundation's Museum.

1978, August

A reproduction of the famous Laird Super Solution is built and returns to the EAA Fly-In Convention where its original pilot, General Jimmy Doolittle, and its original builder, Matty Laird, meet again to recreate their historic participation in the golden age of aviation.

1980, August

The EAA and the EAA Aviation Foundation Boards of Directors vote to establish new EAA facilities on the convention site property already owned in Oshkosh, Wisconsin.

1981

The EAA establishes the first exclusive ultralight organization in the history of the United States when its members form the EAA Ultralight Association. Rapidly growing into the largest ultralight organization in the world, the EAA Ultralight Association begins publishing its own monthly magazine, *Ultralight*.

1982, August

The 30th Annual EAA International Fly-In Convention and Sports Aviation Exhibition breaks all previous records by attracting 750,000 spectators and participants and 14,000 aircraft to Wittman Field in Oshkosh, Wisconsin. In addition, over 200 forums, seminars, and workshops were scheduled throughout the eight-day convention covering every aspect of aviation.

1982, August

Construction begins on the new EAA Aviation Center and World Headquarters in Oshkosh, Wisconsin. In addition to housing a large and sophisticated Air Museum, the new facility will include mini-theaters, restoration shops, full technical and historical libraries, a meeting and conference center, and the EAA World Headquarters offices. The outstanding feature of the new Center is "The Tower of Aviation." This brilliantly lit glass tower rises above the center of the building and will serve as a beacon to both auto and aircraft traffic. Full-size aircraft are suspended within the tower (See Fig. 10-4).

1982, September

The first annual EAA Ultralight Convention is hosted by the EAA Aviation Foundation, EAA, and the EAA Ultralight Association. The first ultralight convention is already the world's largest and most comprehensive ultralight event.

1982, December

The EAA Ultralight Association becomes a division of the National Aeronautic Association (NAA), and is authorized to be sole representative on all ultralight matters within the United States of America reporting directly to the Federation Aeronautique Internationale (FAI). The FAI is the international sport aviation governing body which sanctions aviation record attempts. EAA is honored to be NAA's representative on antique and amateur-built aircraft to the FAI. EAA's International Aerobatic Club is NAA's representative for aerobatic flight.

1983, March 15

Paul H. Poberezny is unanimously elected the president of the Commission Internationale des Aeronefs des Construction Amateur Committee (CIACA) for the 1983/1984 term at the Federation Aeronautique Internationale meeting in Paris.

1983, June 8

The EAA Aviation Foundation's WWII Boeing B-17 bomber plays an important role in a true reunion sponsored by the Miller Brewing Company in Milwaukee. The crew members of the bomber *High Life* see each other again for the first time since the end of WWII.

1983, June 10

EAA successfully petitions the FAA for an exemption to FAR Part 103 governing Ultralight operations. The exemption permits flight training in two-place ultralights.

1983, August

The 1983 EAA Convention is the largest and most successful to date. Approximately 800,000 participants and spectators attend the EAA Con-

Fig. 10-4. No visit to EAA Headquarters would be complete without seeing the new EAA Aviation Center and Air Museum. (photo by Jim Koepnick, courtesy EAA)

vention while approximately 14,000 aircraft fly in during the eight-day celebration of flight. A total of 1,521 show planes registered, and nearly 14,000 visitors from 71 different countries are also attracted to the world's largest aviation event. Over 40,000 aviation enthusiasts utilize the EAA campgrounds during the convention. The nonstop EAA activities feature over 300 forums, seminars, and workshops. More than 400 commercial exhibitors displayed their wares.

1983, August

During the EAA Convention, Federal Aviation Adminstrator J. Lynn Helms presents Paul Poberezny with auto fuel STCs for Cessna 120, 140, 180, and 182 type aircraft. Administrator Helms said that he was pleased to present the STCs to the EAA, saying, "This is an excellent example of what can be done when responsible user groups and the government work together."

1983, August

The winners of the great ARV (aircraft recreational vehicle) fly-off at the EAA Convention are Chuck Slusarczyk of CGS Aviation and Mark Brown of San Antonio, Texas. The ARV design contest, which is partially sponsored by the EAA, is the first aircraft design contest in 20 years. The earlier design contest was also sponsored by the EAA.

1983, August

The new EAA Aviation Center and World Headquarters is dedicated on the EAA convention grounds at Wittman Field in Oshkosh, Wisconsin. The dedication of the EAA Aviation Center is the fruition of Paul H. Poberezny's dream. It is the beginning of a new age for sport and recreational aviation.

1983, August

Long-time EAA member Don Taylor returns to the EAA Convention from his record-breaking trip around the North Pole in a single-engine homebuilt Thorp T-18. Upon his arrival, Taylor donates the T-18 to the EAA Aviation Foundation's Museum in the new EAA Aviation Center.

1983, August 29

The EAA Aviation Foundation and a number of other researchers, suppliers, and manufacturers are invited to testify before the House Subcommittee on Transportation, Aviation, and Materials regarding alternative fuels for general aviation. Testifying on behalf of the Foundation, Harry Zeisloft, Technical Director of the Kermit Weeks Flight Research Center, points out the immediate need for an alternative to avgas and the continuing necessity for alternative fuels in the future. The Subcommittee praises the Foundation's auto fuel research program.

1983, September 23

The National Aeronautics and Space Administration awards Paul H. Poberezny its Distinguished Public Service Medal. The medal, which is the highest award which NASA may bestow upon a U.S. citizen who is not an employee of the federal government, cites Poberezny for his "outstanding contributions to aviation as founder, President, and inspiration of the Experimental Aircraft Association."

1983, September 25

Paul Poberezny is awarded the Federation Aeronautique Internationale's Gold Air Medal in ceremonies at Los Angeles. The presentation takes place during the 76th Annual FAI World Conference. The Gold Air Medal, which is FAI's highest honor, cites Poberezny's aeronautic service on a national and international level, his achievements in aviation, as well as his initiative, and devotion to work involvement for the cause of aviation.

1983, November 22

Paul Poberezny calls a meeting of all concerned aviation entities to discuss EAA's Simplified Type Certification Proposal for a primary aircraft, often referred to as an aircraft recreational vehicle (ARV). He says, "I am pleased that so many have recognized, recently, the need for primary aircraft, which EAA has been promoting for over 20 years." Nearly 40 highly qualified representatives of the FAA, aircraft kit manufacturers, engine manufacturers, and other associated aviation groups accepted Poberezny's invitation for this meeting, which was critical for the development of aviation in the future.

1984, January 9

The entire EAA Headquarters and corporate offices of the EAA Aviation Foundation moved to their new facilities in the EAA Aviation Center in Oshkosh, Wisconsin.

1984, January 31

Tom Poberezny, President of the nonprofit EAA Aviation Foundation, announces that the Foundation has established an educational Air Academy for young people, aged 15 through 17. The first session of the EAA Air Academy will be held July 15 through August 4, 1984. Poberezny says, "The EAA Air Academy offers young people the opportunity to meet and work with aviation professionals, while living and learning the arts, sciences, and lore of aviation in both classroom and workshop settings."

1984, February 6

The themes of both EAA 1984 Conventions are announced as, "The Freedom of Flight." For the second year in a row the Annual EAA International Fly-In Convention and Sport Aviation Exhibition is named one of the top 100 tourist attractions on the North American continent.

1984, February 8

At a meeting of the FAI's Ultralight/Microlight Committee in Paris, the EAA supports FAI's call for the World Ultralight Championship. EAA's Ultralight Grand Prix will be modified to be used as the selection criteria for the official United States World Ultralight Championship Team.

1984, March 20

The nonprofit EAA Aviation Foundation receives FAA approval to issue additional auto fuel STCs for a wide range of aircraft including many Continental-powered Aeronca types, many Continental-powered Piper types, and the 8 series of Luscombe aircraft. A total of 101 separate aircraft models are now eligible for EAA's auto fuel STCs.

1984, March 29

The gondola of the first and only manned balloon to cross the Pacific Ocean, the Double Eagle V, arrives at the EAA Aviation Center. It is destined to become one of the EAA Air Museum's "visitor participation" exhibits in the Air Challenger's Gallery.

1984, April

The EAA Aviation Center, since its dedication in July of 1983, has been toured by over 165,000 visitors. The Air Museum has been roundly praised by aviation enthusiasts, aviation publications, and museum professionals the world over (Fig. 10-4).

"One of the nation's most impressive aircraft museums"— *Aviation Magazine.*

"Surely one of the finest indoor aviation displays in the world"—*Flight International.*

"A true EAA mind-blower. There is not a museum on Earth that can teach EAA anything!"—*Air Progress.*

"It is a noble effort, and well worth your visit"— *Flying.*

MEMBERSHIP INFORMATION

Membership fees are $25 per year, which includes *Sport Aviation* magazine. (I personally feel the magazine *alone* is worth more than this.)

Junior membership is available for those under 18, at $15, and also includes *Sport Aviation.*

For the family that flies together there is family membership at $35 per year, including Sport Aviation.

Membership in a special interest division is extra, but will result in receiving a specialized monthly magazine pertaining only to these particular aircraft.

For further information contact:

EAA
Wittman Airfield
Oshkosh, WI 54903

Chapter 11

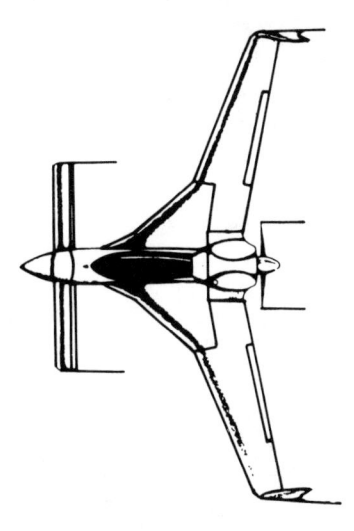

Safety and Hints

The following is a compilation of hints, reminders, and comments to assist the builder in safely completing and flying his project.

PRACTICE KITS

All designers highly recommend that builders who are considering a composite project purchase a practice kit. Wicks Aircraft Supply and Aircraft Spruce & Specialty Company both sell just such a kit. These kits include an assortment of foams, epoxies, fiberglass, filler materials, and supplies with which to work. Everything needed to practice the technique of composite building is included, even Burt Rutan's book *Moldless Composite Sandwich Homebuilt Aircraft Construction*. The cost for this kit is $49.95. I highly recommend it—not just for practice, but to see if you really want to forge ahead and build a complete airplane.

QUALITY CONTROL

One of the unique features of the glass-foam composite construction technique is the ability to visually inspect the structure from the outside. The transparency of the glass/epoxy material makes it possible to see all the way through the skins and even through the spar caps. Defects in the layup take four basic forms:

- ☐ Resin-lean areas.
- ☐ Delaminations.
- ☐ Wrinkles or bumps in the fibers.
- ☐ Damage due to sanding structure away during finishing.

Resin-lean areas are white in appearance due to imcomplete wetting of the glass cloth with epoxy during layup. The presence of minor white (lean) areas up to about two inches in diameter are not cause for rejection of the piece.

Delaminations in a new layup may be due to small air bubbles trapped between plies during the layup. Small delaminations or bubbles up to two-inch diameter may be filled by drilling a small hole into the bubble and filling the void with epoxy.

Major wrinkles or bumps along more than two

inches of chord are cause for rejection in the wings, canard, and winglet on the VariEze, particularly on the top. In most cases the rejected part can be repaired by the same means as a damaged part.

Damaged parts can usually be repaired by following the basic rule: Remove the damaged area and fair back the area at a slope of at least one inch per ply with a sanding block in all directions. Count the number of plies removed while sanding and replace with same, plus one more ply of BID over the entire patch.

HINTS

Your entire work area must not only be spotlessly clean, it must also be warm. Even warming may take three to four hours. This even warming is required by the temperature sensitivity of the epoxies and associated chemicals.

To aid in wetting small areas during cool weather you can use an electric hair drier, being careful not to overheat the part or the epoxy.

Epoxy system components should be stored at room temperature.

Never keep resins or hardeners in a cold place, even for long-term storage.

If the resin appears to crystalize and settle out, it should be returned to its normal state as soon as possible, even if prompt usage is not anticipated. Placing the container of resin in hot (160-190 degrees F.) water for several hours will usually decrystalize and return it to a clear state. Mild agitation will accelerate the process. Leaving the resin hot for three to five hours after it clears will reduce its susceptibility to recrystalize.

Always close containers tightly after use.

Never attempt any layup below 70 degrees since the higher viscosity of the resin will make it more difficult to wet out the cloth. The ideal working temperature is 85 degrees. Keep the epoxy at 75 to 85 degrees. Never work outside in sunlight or in a shop heated with radiant heaters.

Never make a glass layup over a core that is not straight and smooth. The glass panel cannot take the loads if it has bumps or depressions in excess of the allowable values. A wrinkle depression or bump in a layup that is greater than 1/16 inch high or low and which is more than 20 percent of the chord length or 20 percent of the spar chord is not acceptable and requires repair.

Care should be taken in the finishing process not to sand through plies.

A paper cutter is excellent for measuring and cutting the many little squares of glass cloth.

Epoxy should be removed from metal tools or parts with acetone, MEK, or soap and water before it cures.

Micro slurry should not be applied to glass surfaces before being bonded. This weakens the joint.

Do not use Bondo on styrofoam, as it has a polyester content that will dissolve the foam. Bondo will not attack urethane or PVC.

Hot wire cutting: A good method of judging the hot wire temperature is the appearance of the cut foam surface. A cratered or rutted surface indicates the wire is too hot. A very light "hair" of plastic strands on the surface is just right. Always adjust the temperature so that the wire will cut one inch in four to six seconds with light pressure.

Epoxy resins and hardeners are mixed in small batches, usually six ounces or less, even in the largest layups. If mixed in large batches, the heat generated as the harnening progresses speeds the reaction, causing *more* heating, resulting in a very fast reaction called *exotherm*. Exotherm will cause the cup of epoxy to get hot and thicken rapidly. If this occurs, discard it and mix a new batch.

Exotherm foam damage: Care must be taken to avoid heavy buildups of epoxy/micro down inside a joint that is insulated by foam, such as the assembly of the wing cores. The gap to be filled by micro when assembling any foam cores should not be thicker than 1/16 inch. In filling a gap greater than 0.1 inch, excessive weight is added and, more importantly, the large mass of epoxy/micro insulated by the foam can exotherm. Heat resulting from the exotherm can be as high as 450 degrees, which will melt away the foam locally and destroy the joint.

Glass cloth should be stored, marked, and cut in a clean area with clean hands with clean tools. Glass contaminated with dirt, grease, or epoxy should not be used. The area used for storing and cutting glass cloth should be separated from the aircraft assembly area because it will be exposed to foam dust, epoxy, and other elements that can contaminate the cloth.

A pair of good-quality sharp scissors, a felt-tipped marker, a straight board, and a tape measure are needed for marking and cutting. The small amount of ink from marking and numbering plies has no detrimental affect on the glass cloth.

Care and cleanliness should be practiced when working with graphite materials. Unattached graphite fibers are easily airborne. A filter mask should be worn when cutting and fabricating with graphite. Itching and irritation caused by broken filaments becoming embedded in the skin can result from filament breakage during handling.

A simple check of epoxy pot life is to mix about six ounces of epoxy in an eight ounce cup (at 80 degrees F. ambient temperature). Stir well and leave the cup undisturbed. Fast epoxy should exotherm and become solid in 30 to 40 minutes. Slow epoxy will take 80 to 100 minutes for the same result.

Never use wax cups for mixing epoxies, as the wax will contaminate them.

Mixing should be done with wooden mixing sticks.

Use only epoxy with styrofoam. Polyester will dissolve styrofoam.

Either epoxy or polyester systems can be used with urethane and PVC.

Do not store foams in the sunlight.

Do not "hot wire" urethane or PVC foams.

Styrofoam can be hot wired with good ventilation.

At temperatures above 150 degrees F. the plastics used in composite construction can soften, leading to bending—and permanent setting when again cooled. This is one of the reasons for leaving composite airplanes white. White reflects the heat of the sun.

If you must paint decorative stripes on your composite airplane, use light colors. They absorb less heat than dark colors.

HEALTH

Working with all foams can be hazardous to one's health due to inhalation of toxic fumes and dust. All cutting should be done in a well-ventilated area with plenty of fresh air. The use of exhaust fans is recommended and the use of a respirator is absolutely *mandatory*.

Protect yourself from the three dermatitis conditions that can result from contact with composite materials. These are:

- ☐ Contact dermatitis: Normally associated with inflammation of the immediate areas of contact.
- ☐ Irritant dermatitis: Caused by long term contact, causing subcutaneous inflammation.
- ☐ Allergic contact dermatitis: Inflammatory process of the skin (usually acquired over a long period of time and exposure to irritants) resulting in such extreme sensitivity that the mere breathing of fumes can cause a reaction.

There are two methods of protecting one's hands from continued direct contact with irritants. These are barrier creams and gloves.

Barrier creams offer some protection to the user, and facilitate easy cleaning of the hands at the end of a work session. There is no cream that is as effective as a glove.

Gloves offering the best protection are made of butyl rubber and have cotton liners. The rubber offers the protection, and the cotton liner reduces the tendency of the hands to sweat inside the gloves. Gloves made in this manner will last for a long time, and through many work periods. For temporary protection, inexpensive rubber latex gloves are adequate, but not as durable. Vinyl or polyethylene gloves are unacceptable, as they have a high permeability to many chemicals.

PURCHASING

Often you will see partially completed amateur-built airplanes for sale. *Buyer beware*. If you are considering the purchase of a partially completed airplane you should contact the local FAA office (to the airplane) and see what the folks there may know about that *particular* aircraft project. Have an expert evaluate the workmanship for you. Contact your local FAA office and see if you are going to have any problems with the 51 percent rule, and if you will be able to get a Repairman's Certificate when the project is completed.

If you are contemplating the purchase of prebuilt components, check first with your FAA Inspector. It could save many problems later. These inspectors are not there to harass you, they only want to see you complete your project safely . . . *safely!*

FLYING SAFETY

With no reference to written materials, answer the following questions about your airplane:

1. What is the normal climbout speed?
2. What is the best rate-of-climb speed?
3. What is the best angle-of-climb speed?
4. What is the maximum flaps-down speed?
5. What is the maximum gear-down speed?
6. What is the stall speed in normal landing configuration?
7. What is the "clean" stall speed?
8. What is the approach speed?
9. What is the maneuvering speed?
10. What is the never-exceed speed?
11. What engine-out speed will give you maximum gliding range?
12. What is the make and horsepower of your engine?
13. What is the estimated true airspeed at 5,000 feet and 65 percent power?
14. What is your fuel consumption rate at 5,000 feet and 65 percent power?
15. How much usable fuel can you carry?
16. At 5,000 feet and 65 percent what is your duration, less 45 minutes reserve?
17. What type (octane) fuel does your aircraft use?
18. How do you drain the fuel sumps?
19. What weight engine oil and brand is in your engine?
20. What is the maximum crosswind component allowable for your airplane?
21. What is the takeoff distance (over a 50-foot obstacle) at sea level and a temperature of 75 degrees for your airplane?
22. What is the takeoff distance (over a 50-foot obstacle) at 5,000 feet and a temperature of 75 degrees for your airplane?
23. What effect does humidity have on the above numbers?
24. How do you determine pressure altitude?
25. What is your maximum allowable useful load?
26. List two frequencies you can use to contact a FSS on.
27. What is the emergency frequency?

The above questions are based on an AVEMCO safety article. AVEMCO is an aviation insurance company, and provides many safety advisories for the flying public.

Appendices

Appendix A

GADOs and FSDOs

ALABAMA
GADO 2
6500 43rd Avenue, North
Birmingham, AL 35206
Phone: (205) 254-1393

ALASKA
GADO 1
1515 East 13th Avenue
Anchorage, AK 99501
Phone: (907) 272-1324 and 279-5231

FSDO 61
3788 University Avenue
Fairbanks, AK 99701
Phone (907) 452-1276

FSDO 62
Post Office Box 2118
Juneau, AK 99701
Phone: (907) 586-3700

ARIZONA
GADO 9
15041 North Airport Drive
Scottsdale, AZ 85260
Phone: (602) 241-2561

ARKANSAS
GADO 6
FAA NWS Building, Room 201
Adams Field
Little Rock, AR 72202
Phone: (501) 372-3437

CALIFORNIA
GADO 1
7120 Hayvenhurst Avenue, Suite 316
Van Nuys, CA 91406
Phone: (213) 997-3191

GADO 2
1387 Airport Boulevard
San Jose, CA 95110
Phone: (408) 275-7681

GADO 3
3750 John J. Montgomery Drive
San Diego, CA 95110
Phone: (714) 293-5280

GADO 4
Fresno Air Terminal
2401 North Ashley
Fresno, CA 93727
Phone: (209) 487-5306

GADO
Santa Monica Municipal Airport
3200 Airport Avenue, Suite 3
Santa Monica, CA 90405
Phone: (213) 391-6701

GADO 8
Riverside Municipal Airport
6961 Flight Road
Riverside, CA 92504
Phone: (714) 787-1245

GADO 12
Executive Airport
Sacramento, CA 95822
Phone; (916) 440-3169

FSDO 64
P.O. Box 2397
Airport Station
Oakland, CA 94614
Phone: (415) 273-7155

FSDO 65
2815 East Spring Steet
Long Beach, CA 90806
Phone: (213) 426-7134

COLORADO
GADO 3
Jefferson County Airport
Building 1
Broomfield, CO 80020
Phone: (303) 466-7326

GADO 3S
764 Horizon Drive
Grand Junction, CO 81501
Phone: (303) 243-9518

CONNECTICUT
GADO 19
(*See* Massachusetts)

DELAWARE
GADO 9
North Philadelphia Airport
Philadelphia, PA 19114
Phone: (215) 597-9708

DISTRICT OF COLUMBIA
FSDO 62
GT Bldg., Suite 112
Box 17325
Dulles International Airport
Washington, D.C. 20041
Phone: (703) 557-5360

FLORIDA
GADO 5
Building 121
Opa Locka Airport
Opa Locka, FL 33054
Phone: (305) 681-7431

GADO 7
FAA Building
Craig Field
855 Saint John's Bluff Road
Jacksonville, FL 32211
Phone: (904) 641-7311

FSDO 64
Saint Petersburg/Clearwater Airport
Clearwater, FL 33520
Phone: (813) 531-1434

GEORGIA
GADO 1
FAA Building
Fulton County Airport
Atlanta, GA 30336
Phone: (404) 221-6481

HAWAII
FSDO 61
218 Lagoon Drive
Room 215
Honolulu, HI 96819
Phone: (808) 836-0615

IDAHO
GADO 1
3975 Rickenbacker Street
Boise, ID 83705
Phone: (203) 334-1238

ILLINOIS
GADO 3
Post Office Box H
DuPage County Airport
West Chicago, IL 60185
Phone: (312) 584-4490

GADO 19
Capitol Airport
Springfield, IL 62708
Phone: (271) 525-4238

INDIANA
GADO 10
Indianapolis Internation Airport
Box 41525
Indianapolis, IN 46241
Phone: (317) 247-2491

GADO 18
1843 Commerce Drive
South Bend, IN 46628
Phone: (219) 232-5843

IOWA
GADO 4
3021 Army Post Road
Des Moines, IA 50321
Phone: (515) 284-4094

KANSAS
GADO 11
Administration Building
Fairfax Municipal Airport
Kansas City, KS 66115
Phone: (913) 281-3491

GADO 22
Flight Standards Building
Mid-Continent Airport
Wichita, KS 67209
Phone: (316) 943-3244

KENTUCKY
GADO 13
FAA Building
Bowman Field
Louisville, KY 40205
Phone: (502) 582-6116

LOUISIANA
GADO 8
FAA Building
Lakefront Airport
New Orleans, LA 70126
Phone: (504) 241-2506

GADO 8 South
FAA Office
Lafayette Airport
Lafayette, LA 70508
Phone: (318) 234-2321

GADO 11
Terminal Building
Room 137
Downtown Airport
Shreveport, LA 71107
Phone: (318) 226-5379

MAINE
GADO 15
Portland International Jetport
Portland, ME 04102
Phone: (207) 774-4484

MARYLAND
GADO 21
Elm Road
BWI International Airport
Baltimore, MD 21240
Phone: (301) 761-2610

MASSACHUSETTS
GADO 13
Norwood Municipal Airport
Norwood, MA 02062
Phone: (617) 762-2436

GADO 19
Barnes Municipal Airport
Westfield, MA 01085
Phone: (413) 568-3121

MICHIGAN
GADO 8
Kent County International Airport
5500 44th Street SE
Grand Rapids, MI 49508
Phone: (616) 456-2427

GADO 20
Flight Standards Building
Willow Run Airport
Box 860
Ypsilanti, MI 48197
Phone: (313) 485-2550

MINNESOTA
GADO 14
6201 34th Avenue South
Minneapolis, MN 55450
Phone: (612) 725-3341

MISSISSIPPI
GADO 4
FAA Building Municipal Airport
Box 6273
Pearl Branch
Jackson, MS 39208
Phone: (601) 969-4633

MISSOURI
FSDO 62
9275 Jenaire Drive
Burkley, MO 63134
Phone: (314) 425-7100

MONTANA
FSDO 61
Administration Building
Room 216
Billings Logan International Airport
Billings, MT 59101
Phone: (406) 245-6179

FSDO 65
FAA Building
Room 3
Helena Airport
Helena, MT 59601
Phone: (406) 499-5270

NEBRASKA
GADO 12
General aviation Building
Lincoln Municipal Airport
Lincoln, NE 68521
Phone: (402) 471-5485

NEVADA
GADO 11
601 South Rock Blvd.
Suite 102
Reno, NV 89502
Phone: (702) 784-5321
FSDO 66
5700 C South Haven
Las Vegas, NV 89119
Phone: (702) 736-0666

NEW HAMPSHIRE
GADO 15
(*See* Maine)

NEW JERSEY
FSDO 61
150 Riser Road
Teterboro Airport
Teterboro, NJ 07608
Phone: (201) 288-1745

NEW MEXICO
GADO 1
2402 Kirtland Drive, SE
Albuquerque, NM 87106
Phone: (505) 247-0156

NEW YORK
GADO 1
Albany County Airport
Albany, NY 12211
Phone: (518) 869-8482

GADO 11
Building 53
Republic Airport
Farmingdale, NY 11735
Phone: (516) 694-5530

GADO 17
Rochester-Monroe Airport
Rochester, NY 14624
Phone: (716) 263-5880

NORTH CAROLINA
GADO 3
FAA Building
Box 27005
Charlotte, NC 28219
Phone: (704) 392-3214

GADO 11
Route 1, Box 486A
Morrisville, NC 27560
Phone: (919) 755-4240

NORTH DAKOTA
FSDO 64
Box 5496
Fargo, ND 58105
Phone: (701) 232-8949

OHIO
GADO 5
4242 Airport Road
Lunken Executive Building
Cincinnati, OH 45226
Phone: (513) 684-2183

GADO 6
Federal Facilities Building
Cleveland Hopkins International Airport
Cleveland, OH 44135
Phone: (216) 267-0220

GADO 7
4393 East 17th Avenue
Port Columbus International Airport
Columbus, OH 43219
Phone: (614) 469-7476

OKLAHOMA
GADO 9
FAA Building
Wiley Post Airport
Bethany, OK 73008
Phone: (405) 789-5220

FSDO 65
General Aviation Terminal Building
Room 103
Tulsa International Airport
6501 E Apache
Tulsa, OK 74115
Phone: (918) 835-2378

OREGON
GADO 2
Mahlon Sweet Airport
90606 Greenhill Road
Eugene, OR 97402
Phone: (503) 688-9721

GADO 3
Portland/Hillsboro Airport
3355 NE Cornell Road
Hillsboro, OR 97123
Phone: (503) 221-2104

PENNSYLVANIA
GADO 3
Allentown-Bethlehem-Easton Airport
Allentown, PA 18103
Phone: (215) 264-2888

GADO 9
North Philadelphia Airport
Philadelphia, PA 19114
Phone: (215) 597-9708

GADO 10
Room 201
Administration Building
Capitol City Airport
New Cumberland, PA 17070
Phone: (717) 782-4528

GADO 14
Allegheny County Airport
West Mifflin, PA 15122
Phone: (412) 462-5507

RHODE ISLAND
GADO 13
(*See* Massachusetts)

SOUTH CAROLINA
GADO 9
Columbia Metropolitan Airport
West Columbia, SC 29169
Phone: (803) 765-5931

SOUTH DAKOTA
FSDO 66
Rural Route 2, Box 633B
Rapids City, SD 57701
Phone: (605) 343-2403

TENNESSEE
FSDO 62
322 Knapp Blvd.
Room 101
Nashville Metropolitan Airport
Nashville, TN 37217
Phone: (615) 251-5661

FSDO 63
2488 Winchester
Room 137
Memphis, TN 38116
Phone: (901) 345-0600

TEXAS
GADO 2
8032 Aviation Place
Love Field Airport
Dallas, TX 75235
Phone: (214) 357-0142

GADO 3
FAA NWS Building
Room 202
6795 Convair Road
El Paso, TX 79925
Phone: (915) 778-6389

FSDO 61
Administration Building
Room 240
Meacham Field
Fort Worth, TX 76106
Phone: (817) 624-4911

FSDO 62
8800 Paul B Koonce Drive
Room 224
Houston, TX 77061
Phone: (713) 645-6628

GADO 7
Route 3, Box 51
Lubbock, TX 79401
Phone: (806) 762-0335

FSDO 64
1115 Paul Wilkins Road
Room 201
San Antonio, TX 78216
Phone: (512) 824-9535

UTAH
FSDO 67
116 North 2400 West
Salt Lake City, UT 84116
Phone: (801) 524-4247

VERMONT
GADO 15
(*See* Maine)

VIRGINIA
GADO 16
Byrd Field
Sandstone, VA 23150
Phone: (804) 222-7494

FSDO 62
GT Building, Suite 112
Box 17325
Dulles International Airport
Washington, D.C. 20041
Phone: (703) 557-5360

WASHINGTON
GADO 5
5620 East Rutter Avenue
Spokane, WA 99206
Phone: (509) 456-4618

FSDO 61
FAA Building
Boeing Field
Seattle, WA 98108
Phone: (206) 767-2724

WEST VIRGINIA
GADO 22
301 Eagle Mountain Road
Room 144
Kanawha Airport
Charleston, WV 25311
Phone: (304) 343-4689

WISCONSIN
FSDO 61
General Mitchell Field
FAA/WB Building
Milwaukee, WI 53207
Phone: (414) 747-5531

WYOMING
FSDO 62
Natrona County International Airport
FAA/WB Building
Casper, WY 82601
Phone: (307) 234-8959

Appendix B

Manufacturers, Suppliers, and Organizations

AeroMotion Inc.
1224 W. South Park Ave.
Oshkosh, WI 54901
Phone: (414) 233-0773

Aircraft Owners and Pilots Association
421 Aviation Way
Frederick, MD 21701
Phone: (301) 695-2000

Aircraft Spruce & Specialty Co.
P.O. Box 424
Fullerton, CA 92632
Phone: (714) 870-7551

Air Pix Aviation Photography
P.O. Box 75034
Cincinnati, OH 45275

Antenna Dynamics Inc.
1251 W. Sepulveda Blvd.
Suite #268
Torrance, CA 90502
Phone: (213) 534-1090 ext 22

APCO (Applied Plastics Co., Inc.)
612 East Franklin Ave.
El Segundo, CA 90245

Avco Lycoming
Williamsport, PA 17701
Phone: (717) 323-6181

Communications Specialists, Inc.
426 Taft Ave.
Orange, CA 92665
Phone: (800) 854-0547

Davtron
427 Hillcrest Way
Redwood City, CA 94062
Phone: (415) 369-1188

Duncan Aviation Engines
Rt. 1 Box 256
Comanche, OK 73529
Phone: (405) 439-2473

Experimental Aircraft Association
Wittman Airfield
Oshkosh, WI 54903
Phone: (414) 426-4800

Great American Propeller Co.
1180 Pike Lane #5
Oceano, CA 93445
Phone: (805) 481-9054

Great Plains Aircraft Supply Co., Inc.
P.O. Box 1481
Palatine, IL 60078
Phone: (312) 359-6558

HAPI Engines, Inc.
RR 1 Box 1000
Eloy, AZ 85231
Phone: (602) 466-9244

(Monnett Experimental Aircraft, Inc.)
INAV Ltd.
P.O. Box 2984
Oshkosh, WI 54903
Phone: (414) 426-1212

Ken Brock Manufacturing Co.
11852 Western Ave.
Stanton, CA 90680
Phone: (714) 898-4366

King Radio Corp.
400 N Rogers Rd
Olathe, KS 66062
Phone: (913) 782-0400

Limbach Aircraft Engines
P.O. Box 1201
Tulsa, OK 74101
Phone: (918) 245-6910

Narco Avionics
270 Commerce Dr
Ft. Washington, PA 19034
Phone: (215) 643-2900

Radio Systems Technology
13281 Grass Valley Ave.
Grass Valley, CA 95945
Phone: (916) 272-2203

Sensenich Corp.
Airport Road
P.O. Box 4187
Lancaster, PA 17604

TASK Research, Inc.
848 East Santa Maria St.
Santa Paula Airport
Santa Paula, CA 93060
Phone: (805) 525-4445

Telex Communications, Inc.
9600 Aldrich Ave. South
Minneapolis, MN 55420
Phone: (612) 884-4051

Teledyne Continental Motors
P.O. Box 90
Mobile, AL 36601
Phone: (205) 438-3411

Terra Corp.
3520 Pan American Freeway NE
Albuquerque, NM 87107
Phone: (505) 884-2321

Trade-A-Plane
Crossville, TN 38555

Wicks Aircraft Supply
410 Pine
Highland, IL 62249

Appendix C

Construction Measurements

The following tables, formulas, parts lists, and templates are provided by Aircraft Spruce and Specialty Company to assist the planner/builder along the path to a completed aircraft.

TAP DRILL SIZES

N.F. (S.A.E.)			N.C. (U.S.S.)			MACHINE SCREW				AMER. Std. Pipe		
Diameter of Tap	Threads Per Inch	Size of Drill	Diameter of Tap	Threads Per Inch	Size of Drill	Number Sizes of Taps	Threads Per Inch	Tap DRILLS	Body or Clearance	Diameter of Tap	Threads Per Inch	Size of Drill (BRIGGS STANDARD)
¼	28	No. 3	¼	20	No. 7	2	56	No. 50	No. 43	⅛	27	11/32
5/16	24	I	5/16	18	F	3	48	No. 46	No. 40	¼	18	7/16
⅜	24	Q	⅜	16	5/16	4	40	No. 43	No. 32	⅜	18	37/64
7/16	20	25/64	7/16	14	U	6	32	No. 35	No. 28	½	14	23/32
½	20	29/64	½	13	27/64	8	32	No. 29	No. 19	¾	14	59/64
9/16	18	33/64	9/16	12	31/64	10	24	No. 25	No. 10	1	11½	1-5/32
⅝	18	37/64	⅝	11	17/32	10	32	No. 21	No. 10	1-¼	11½	1-½
11/16	16	⅝	11/16	11	19/32	12	24	No. 16	7/32	1-½	11½	1-47/64
¾	16	11/16	¾	10	21/32	FRACTIONAL SIZES UNDER ¼"				2	11½	2-7/32
13/16	14	13/16	13/16	10	23/32	1/16	64	3/64		2-½	8	2-5/8
⅞	14	15/16	⅞	9	49/64	3/32	48	No. 49		3	8	3-¼
1	14	15/16	1	9	53/64	⅛	40	No. 38		3-½	8	3-¾
1-⅛	12	1-3/64	15/16	8	⅞	5/32	32	⅛		4	8	4-¼
1-¼	12	1-11/64	1	8	63/64	3/16	36	No. 30				
1-½	12	1-27/64	1-⅛	7	1-7/64	3/16	24	No. 26				
			1-¼	7	1-7/64	¼	32	No. 22				
			1-½	6	1-11/32	7/32	24	No. 16				
						7/32	32	No. 12				

THE TAP DRILL SIZES ABOVE ARE FOR USE WITH PLUG TAPS—SAVING THE NECESSITY FOR TAPER OR STARTING TAPS. THE DRILL SIZES ARE FOR USE WITH TAPER PIPE TAPS

147

CENTIGRADE–FAHRENHEIT CONVERSION TABLE

Locate Temperature to be converted in center column Read Centigrade (celsius) equivalents to left, Fahrenheit equivalents to right

C	F or C	F	C	F or C	F	C	F or C	F	C	F or C	F
−56.7	−70	−93	16.1	61	141.8	52.4	126	258.8	88.5	191	375.8
−51.1	−60	−76	16.7	62	143.6	52.9	127	260.6	89.0	192	377.6
−45.6	−50	−58	17.2	63	145.4	53.5	128	262.4	89.6	193	379.4
−40.0	−40	−40	17.8	64	147.2	54.0	129	264.2	90.1	194	381.2
−34.4	−30	−22	18.3	65	149	54.5	130	266	90.7	195	383
−28.9	−20	−4	18.9	66	150.8	55.1	131	267.8	91.3	196	384.8
−23.3	−10	14	19.4	67	152.6	55.6	132	269.6	91.8	197	386.6
−17.8	0	32	20.0	68	154.4	56.2	133	271.4	92.4	198	388.4
−17.2	1	33.8	20.6	69	156.2	56.7	134	273.2	92.9	199	390.2
−16.7	2	35.6	21.1	70	158	57.3	135	275	93.4	200	392
−16.1	3	37.4	21.7	71	159.8	57.8	136	276.8	94.0	201	393.8
−15.6	4	39.2	22.2	72	161.6	58.3	137	278.6	94.5	202	395.6
−15.0	5	41	22.8	73	163.4	58.9	138	280.4	95.1	203	397.4
−14.4	6	42.8	23.3	74	165.2	59.4	139	282.2	95.6	204	399.2
−13.9	7	44.6	23.9	75	167	60.0	140	284	96.2	205	401.
−13.3	8	46.4	24.4	76	168.8	60.6	141	285.8	96.8	206	402.8
−12.8	9	48.2	25.0	77	170.6	61.1	142	287.6	97.3	207	404.6
−12.2	10	50	25.6	78	172.4	61.7	143	289.4	97.9	208	406.4
−11.7	11	51.8	26.1	79	174.2	62.2	144	291.2	98.4	209	408.2
−11.1	12	53.6	26.7	80	176	62.8	145	293	99.0	210	410
−10.6	13	55.4	27.2	81	177.8	63.4	146	294.8	99.5	211	411.8
−10.0	14	57.2	27.8	82	179.6	63.9	147	296.6	100.0	212	413.6
−9.44	15	59	28.3	83	181.4	64.5	148	298.4	100.6	213	415.4
−8.89	16	61.8	28.9	84	183.2	65.0	149	300.2	101.1	214	417.2
−8.33	17	63.6	29.4	85	185	65.6	150	302	101.7	215	419
−7.78	18	65.4	30.0	86	186.8	66.1	151	303.8	102.2	216	420.8
−7.72	19	67.2	30.6	87	188.6	66.6	152	305.6	102.8	217	422.6
−6.67	20	68	31.1	88	190.4	67.2	153	307.4	103.3	218	424.4
−6.11	21	69.8	31.7	89	192.2	67.7	154	309.2	103.9	219	426.2
−5.56	22	71.6	32.2	90	194	68.3	155	311	104.4	220	428
−5.00	23	73.4	32.8	91	195.8	68.9	156	312.8	105.0	221	429.8
−4.44	24	75.2	33.3	92	197.6	69.4	157	314.6	105.6	222	431.6
−3.89	25	77	33.9	93	199.4	70.0	158	316.4	106.1	223	433.4
−3.33	26	78.8									

C	F	C	F	C	F	C	F	C	F
−2.78	27	80.6	34.4	94	201.2	70.5	159	318.2	106.7	224	435.2
−2.22	28	82.4	35.0	95	203	71.1	160	320	107.2	225	437
−1.67	29	84.2	35.6	96	204.8	71.7	161	321.8	107.8	226	438.8
−1.11	30	86	36.1	97	206.6	72.2	162	323.6	108.3	227	440.6
−0.56	31	87.8	36.7	98	208.4	72.8	163	325.4	108.9	228	442.4
0	32	89.6	37.2	99	210.2	73.3	164	327.2	109.4	229	444.2
0.56	33	91.4	37.8	100	212	73.9	165	329	110.0	230	446
1.11	34	93.2	38.4	101	213.8	74.5	166	330.8	110.6	231	447.8
1.67	35	95	38.9	102	215.6	75.0	167	332.6	111.1	232	449.6
2.22	36	96.8	39.5	103	217.4	75.6	168	334.4	111.7	233	451.4
2.78	37	98.6	40.0	104	219.2	76.1	169	336.2	112.2	234	453.2
3.33	38	100.4	40.6	105	221	76.7	170	338	112.8	235	455
3.89	39	102.2	41.2	106	222.8	77.3	171	339.8	113.3	236	456.8
4.44	40	104	41.7	107	224.6	77.8	172	341.6	113.9	237	458.6
5.00	41	105.8	42.3	108	226.4	78.4	173	343.4	114.4	238	460.4
5.56	42	107.6	42.8	109	228.2	78.9	174	345.2	115.0	239	462.2
6.11	43	109.4	43.4	110	230	79.5	175	347	115.5	240	464
6.67	44	111.2	44.0	111	231.8	80.1	176	348.8	116.1	241	465.8
7.22	45	113	44.5	112	233.6	80.6	177	350.6	116.6	242	467.6
7.78	46	114.8	45.1	113	235.4	81.2	178	352.4	117.2	243	469.4
8.33	47	116.6	45.6	114	237.2	81.7	179	354.2	117.7	244	471.2
8.89	48	118.4	46.2	115	239	82.3	180	356	118.3	245	473
9.44	49	120.2	46.8	116	240.8	82.9	181	357.8	118.9	246	474.8
10.0	50	122	47.3	117	242.6	83.4	182	359.6	119.4	247	476.6
10.6	51	123.8	47.9	118	244.4	84.0	183	361.4	120.0	248	478.4
11.1	52	125.6	48.4	119	246.2	84.5	184	363.2	120.5	249	480.2
11.7	53	127.4	49.0	120	248	85.1	185	365	121.1	250	482
12.2	54	129.2	49.6	121	249.8	85.7	186	366.8
12.8	55	131	50.1	122	251.6	86.2	187	368.6
13.3	56	132.8	50.7	123	253.4	86.8	188	370.4
13.9	57	134.6	51.2	124	255.2	87.3	189	372.2
14.4	58	136.4	51.8	125	257	87.9	190	374
15.0	59	138.2
15.6	60	140

For unlisted temperatures, use following formulas:

$$C = \frac{5F - 160}{9} \qquad F = \frac{9C + 160}{5}$$

LENGTH EQUIVALENTS

Unit	Milli-meters	Centi-meters	Inches	Feet	Yards	Meters
1 Millimeter =	1	.1	.03937	.003281	.001094	.001
1 Centimeter =	10	1	.3937	.032808	.010936	.01
1 Inch =	25.4001	2.54001	1	.083333	.027778	.025400
1 Foot =	304.801	30.4801	12	1	.333333	.304801
1 Yard =	914.402	91.4402	36	3	1	.914402
1 Meter =	1000	100	39.37	3.28083	1.09361	1

Unit	Feet	Yards	Meters	Rods	Furlongs	Miles (Statute)
1 Rod =	16.5	5.5	5.02921	1	.025 (1/40)	.003125 (1/320)
1 Furlong =	660	220	201.168	40	1	.125 (1/8)
1 Mile (statute) =	5280	1760	1609.35	320	8	1

1 Nautical Mile = 6080.2 feet = 1.15155 statute miles = ⅓ league.
1 Light Year = 5.879 trillion miles = 9.46 trillion kilometers.

WEIGHT EQUIVALENTS

Unit	Grains	Grams	Ounces (Troy)	Ounces (Avoir.)	Pounds (Troy)	Pounds (Avoir.)	Kilo-grams
1 Grain =	1	.064799	.002083	.002286	.000174	.000143	.000065
1 Gram =	15.4324	1	.032151	.035274	.002679	.002205	.001
1 Ounce (Troy) =	480	31.1035	1	1.09714	.083333	.068571	.031104
1 Ounce (Avoir.) =	437.5	28.3495	.911458	1	.075955	.0625	.028350
1 Pound (Troy) =	5760	373.242	12	13.1657	1	.822857	.373242
1 Pound (Avoir.) =	7000	453.592	14.5833	16	1.21528	1	.453592
1 Kilogram =	15432.4	1000	32.1507	35.2740	2.67923	2.20462	1

Unit	Kilograms	Pounds (Troy)	Pounds (Avoir.)	Metric Tons	Net (Short) Tons	Gross (Long) Tons
1 Metric Ton =	1000	2679.23	2204.62	1	1.10231	.984206
1 Net (Short) Ton =	907.185	2430.56	2000	.907185	1	.892857
1 Gross (Long) Ton =	1016.05	2722.22	2240	1.01605	1.12	1

METRIC SYSTEM

LENGTH

1 meter (m) = { 10 decimeters (dm); 100 centimeters (cm); 1,000 millimeters (mm) }

1 dekameter (dkm) = 10 meters (m)
1 hectometer (hm) = 100 meters (m)
1 kilometer (km) = 1,000 meters (m)

WEIGHT

1 gram (g) = { 10 decigrams (dg); 100 centigrams (cg); 1,000 milligrams (mg) }

1 dekagram (dkg) = 10 grams (g)
1 hectogram (hg) = 100 grams (g)
1 kilogram (kg) = 1,000 grams (g)

1 metric ton = { 1,000 kilograms (kg); 1,000,000 grams (g) }

VOLUME & CAPACITY

1 liter (l) = { 1 cubic decimeter (dm^3); 10 deciliters (dl); 100 centiliters (cl); 1,000 milliliters (ml); 1,000 cubic centimeters (cm^3 or cc) }

1 dekaliter (dkl) = 10 liters (l)
1 hectoliter (hl) = 100 liters (l)

1 kiloliter (kl) = { 1 cubic meter (m^3); 1 stere (s); 1,000 liters (l) }

VOLUME AND CAPACITY EQUIVALENTS

Unit	Cubic Centimeters	Cubic Inches	Liters	Quarts (Liquid)	Quarts (Dry)	Gals. (Liquid)	Gals. (Dry)	Cubic Feet
1 Cu. Centimeter =	1	.06102	.001	.00106	.00091	.00026	.00023	.00004
1 Cu. Inch =	16.387	1	.01639	.01732	.01488	.00433	.00372	.00058
1 Gill =	118.29	7.2188	.11829	.125	.10742	.03125	.02686	.00418
1 Pint (liquid) =	473.18	28.875	.47318	.5	.42968	.125	.10742	.01671
1 Pint (dry) =	550.62	33.600	.55062	.58182	.5	.14546	.125	.01945
1 Liter =	1000	61.023	1	1.0567	.90808	.26417	.22702	.03531
1 Quart (liquid) =	946.36	57.75	.94636	1	.85937	.25	.21484	.03342
1 Quart (dry) =	1101.2	67.201	1.1012	1.1637	1	.29091	.25	.03889
1 Gallon (liquid) =	3785.4	231	3.7854	4	3.4375	1	.85937	.13368
1 Gallon (dry) =	4404.9	268.80	4.4049	4.6546	4	1.1636	1	.15556
1 Peck =	8809.8	537.81	8.8098	9.3092	8	2.3273	2	.31111
1 Cu. Foot =	28317.0	1728	28.317	29.922	25.714	7.4805	6.4285	1
1 Bushel =	35239.3	2150.4	35.239	37.237	32	9.3092	8	1.2445
1 Barrel =	119241.2	7278.5	119.24	126	108.28	31.5	27.070	4.2109
1 Cu. Yard =	764559.4	46656	764.56	807.90	694.28	201.97	173.57	27
1 Cu. Meter =	1000000	61023.4	1000	1056.7	908.08	264.17	227.02	35.314

AREA EQUIVALENTS

Unit	Square Inches	Square Feet	Square Yards	Square Meters
1 Square Foot =	144	1	.1111	.09290
1 Square Yard =	1296	9	1	.83613
1 Square Meter =	1550	10.7639	1.19599	1
1 Square Rod =	39204	272.25	30.25	25.293
1 Are =	155000	1076.39	119.599	100
1 Acre =	6272640	43560	4840	4046.86
1 Square Mile (640 Acres) =	—	27878400	3097600	2589999
1 Square Kilometer =	—	10763867	1195985	1000000

MISCELLANEOUS LIQUID MEASURE CONVERSIONS

1 Fl. Oz.	=	1.805 Cu. In.	1 Quart	=	946 Cu. Cm.
1 Fl. Oz.	=	29.6 Cu. Cm.	1 Gallon	=	8 Pints
1 Pint	=	16 Fl. Oz.	1 Gallon	=	4 Quarts
1 Quart	=	32 Fl. Oz.	1 Firkin	=	9 Gallons
1 Quart	=	2 Pints	1 Hogshead	=	63 Gallons

AREA

1 centare (ca) = 1 square meter (m^2); 100 square decimeters (dm^2); 10,000 square centimeters (cm^2); 1,000,000 square millimeters (mm^2)

1 are (a) = 1 square dekameter (dkm^2); 100 square meters (m^2)

1 hectare (ha) = 100 ares (a); 10,000 square meters (m^2)

1 square kilometer (km^2) = 1,000,000 square meters (m^2)

Other prefixes occasionally used:

micro — one millionth
deca — 10 times (same as deka)
myria — 10,000 times
mega — 1,000,000 times

AN/MS O-RING EQUIVALENTS

The AN6227B, AN6230B and MS28775 O-Rings have the same composition. For use in aircraft hydraulic systems using MIL-H-5606 petroleum base hydraulic fluid.

AN6227B/MS28775 Equivalents

AN6227B Dash No.	MS28775 Dash No.	Nominal Size Width	Nominal Size I.D.	Nominal Size O.D.
1	6	1/16	1/8	1/4
2	7	1/16	5/32	9/32
3	8	1/16	3/16	5/16
4	9	1/16	7/32	11/32
5	10	1/16	1/4	3/8
6	11	1/16	5/16	7/16
7	12	1/16	3/8	1/2
8	110	3/32	3/8	9/16
9	111	3/32	7/16	5/8
10	112	3/32	1/2	11/16
11	113	3/32	9/16	3/4
12	114	3/32	5/8	13/16
13	115	3/32	11/16	7/8
14	116	3/32	3/4	15/16
15	210	1/8	3/4	1
16	211	1/8	13/16	1-1/16
17	212	1/8	7/8	1-1/8
18	213	1/8	15/16	1-3/16
19	214	1/8	1	1-1/4
20	215	1/8	1-1/16	1-5/16
21	216	1/8	1-1/8	1-3/8
22	217	1/8	1-3/16	1-7/16
23	218	1/8	1-1/4	1-1/2
24	219	1/8	1-5/16	1-9/16
25	220	1/8	1-3/8	1-5/8
26	221	1/8	1-7/16	1-11/16
27	222	1/8	1-1/2	1-3/4
28	325	3/16	1-1/2	1-7/8
29	326	3/16	1-5/8	2
30	327	3/16	1-3/4	2-1/8

AN6230B/MS28775 Equivalents

AN6230B Dash No.	MS28775 Dash No.	Nominal Size Width	Nominal Size I.D.	Nominal Size O.D.
1	223	1/8	1-5/8	1-7/8
2	224	1/8	1-3/4	2
3	225	1/8	1-7/8	2-1/8
4	226	1/8	2	2-1/4
5	227	1/8	2-1/8	2-3/8
6	228	1/8	2-1/4	2-1/2
7	229	1/8	2-3/8	2-5/8
8	230	1/8	2-1/2	2-3/4
9	231	1/8	2-5/8	2-7/8
10	232	1/8	2-3/4	3
11	233	1/8	2-7/8	3-1/8
12	234	1/8	3	3-1/4
13	235	1/8	3-1/8	3-3/8
14	236	1/8	3-1/4	3-1/2
15	237	1/8	3-3/8	3-5/8
16	238	1/8	3-1/2	3-3/4
17	239	1/8	3-5/8	3-7/8
18	240	1/8	3-3/4	4
19	241	1/8	3-7/8	4-1/8
20	242	1/8	4	4-1/4
21	243	1/8	4-1/8	4-3/8
22	244	1/8	4-1/4	4-1/2
23	245	1/8	4-3/8	4-5/8
24	246	1/8	4-1/2	4-3/4
25	247	1/8	4-5/8	4-7/8

No MS28775 Equivalent for AN6230B-31 and Larger

HOW TO USE THIS CHART

The table of wood screw specifications simplifies the selection of the bit or drill size best suited to your requirements. The fractional equivalents and undersize and oversize decimals indicate how close a bit of given fractional size will bore to the actual screw dimension and whether the fit will be snug or loose. In selecting a tool size for the pilot hole (for threaded portion of screw), note that root diameters are average dimensions measured at the middle of the threaded portion. On some screws the root diameter tapers slightly from the end of the screw, increasing toward the head. It is usually good practice to bore the pilot hole the same size as the root diameter in hardwoods, such as oak, and about 15% smaller for soft woods, such as pine and Douglas fir. In some cases, allowances can be made to advantage for moisture content and other varying factors. This same rule can be used for shank holes. The SHANK DIAMETERS shown below are standard specifications subject to tolerances of +.004/-.007. MAXIMUM HEAD DIAMETERS are also standard specifications which apply to flat and oval-head screws. Head sizes run from 5% to 10% smaller for round-head screws.

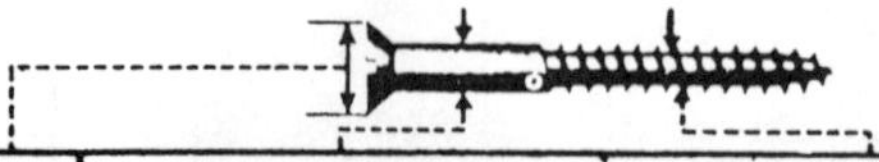

NO. OF SCREW	MAXIMUM HEAD DIAMETER	SHANK DIAMETER BASIC DEC. SIZE	SHANK DIAMETER NEAREST FRACTIONAL EQUIVALENT	ROOT DIAMETER AVERAGE DEC. SIZE	ROOT DIAMETER NEAREST FRACTIONAL EQUIVALENT	THREADS PER INCH	NO. OF SCREW
0	.119	.060	1/16 OVERSIZE .002	.040	3/64 OVERSIZE .007	32	0
1	.146	.073	5/64 OVERSIZE .005	.046	3/64 BASIC SIZE	28	1
2	.172	.086	3/32 OVERSIZE .007	.054	1/16 OVERSIZE .008	26	2
3	.199	.099	7/64 OVERSIZE .010	.065	1/16 UNDERSIZE .002	24	3
4	.225	.112	7/64 UNDERSIZE .003	.075	5/64 OVERSIZE .003	22	4
5	.252	.125	1/8 BASIC SIZE	.085	5/64 UNDERSIZE .007	20	5
6	.279	.138	9/64 OVERSIZE .002	.094	3/32 BASIC SIZE	18	6
7	.305	.151	5/32 OVERSIZE .005	.102	7/64 OVERSIZE .007	16	7
8	.332	.164	5/32 UNDERSIZE .007	.112	7/64 UNDERSIZE .003	15	8
9	.358	.177	11/64 UNDERSIZE .005	.122	1/8 OVERSIZE .003	14	9
10	.385	.190	3/16 UNDERSIZE .002	.130	1/8 UNDERSIZE .005	13	10
11	.411	.203	13/64 BASIC SIZE	.139	9/64 OVERSIZE .001	12	11
12	.438	.216	7/32 OVERSIZE .003	.148	9/64 UNDERSIZE .007	11	12
14	.491	.242	1/4 OVERSIZE .008	.165	5/32 UNDERSIZE .009	10	14
16	.544	.268	17/64 UNDERSIZE .002	.184	3/16 OVERSIZE .003	9	16
18	.597	.294	19/64 OVERSIZE .003	.204	13/64 UNDERSIZE .001	8	18
20	.650	.320	5/16 UNDERSIZE .007	.223	7/32 UNDERSIZE .004	8	20
24	756	.372	3/8 OVERSIZE .003	.260	1/4 UNDERSIZE .010	7	24

ACTUAL WOOD SCREW SHANK SIZES

To determine the size of a screw visually, lay the screw shank on the silhouette.

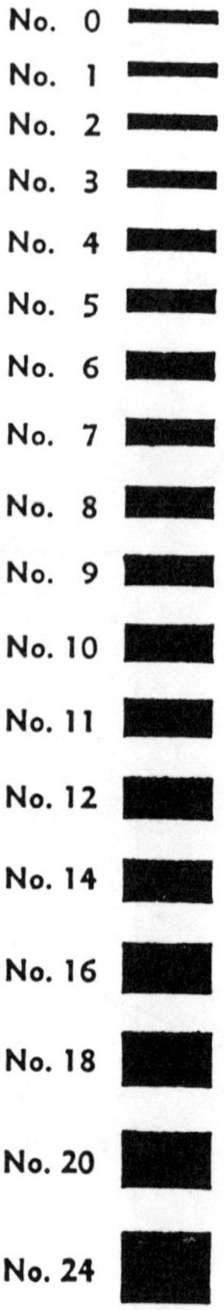

No. 0
No. 1
No. 2
No. 3
No. 4
No. 5
No. 6
No. 7
No. 8
No. 9
No. 10
No. 11
No. 12
No. 14
No. 16
No. 18
No. 20
No. 24

FRACTIONAL INCHES CONVERTED TO DECIMAL INCHES AND MILLIMETERS

Fraction of Inch	Decimal Inch	Decimal Millimeters	Fraction of Inch	Decimal Inch	Decimal Millimeters	Fraction of Inch	Decimal Inch	Decimal Millimeters	Fraction of Inch	Decimal Inch	Decimal Millimeters
1/64	.015625	0.39688	17/64	.265625	6.74689	33/64	.515625	13.09690	49/64	.765625	19.44691
1/32	.03125	0.79375	9/32	.28125	7.14376	17/32	.53125	13.49378	25/32	.78125	19.84379
3/64	.046875	1.19063	19/64	.296875	7.54064	35/64	.546875	13.89065	51/64	.796875	20.24067
1/16	.0625	1.58750	5/16	.3125	7.93752	9/16	.5625	14.28753	13/16	.8125	20.63754
5/64	.078125	1.98438	21/64	.328125	8.33439	37/64	.578125	14.68440	53/64	.828125	21.03442
3/32	.09375	2.38125	11/32	.34375	8.73127	19/32	.59375	15.08128	27/32	.84375	21.43129
7/64	.109375	2.77813	23/64	.359375	9.12814	39/64	.609375	15.47816	55/64	.859375	21.82817
1/8	.125	3.17501	3/8	.375	9.52502	5/8	.625	15.87503	7/8	.875	22.22504
9/64	.140625	3.57188	25/64	.390625	9.92189	41/64	.640625	16.27191	57/64	.890625	22.62192
5/32	.15625	3.96876	13/32	.40625	10.31877	21/32	.65625	16.66878	29/32	.90625	23.04880
11/64	.171875	4.36563	27/64	.421875	10.71565	43/64	.671875	17.06566	59/64	.921875	23.41567
3/16	.1875	4.76251	7/16	.4375	11.11252	11/16	.6875	17.46253	15/16	.9375	23.81255
13/64	.203125	5.15939	29/64	.453125	11.50940	45/64	.703125	17.85941	61/64	.953125	24.20942
7/32	.21875	5.55626	15/32	.46875	11.90627	23/32	.71875	18.25629	31/32	.96875	24.60630
15/64	.234375	5.95314	31/64	.484375	12.30315	47/64	.734375	18.65316	63/64	.984375	25.00318
1/4	.25	6.35001	1/2	.5	12.70003	3/4	.75	19.05004	1	1.	25.40005

MILLIMETERS CONVERTED TO DECIMAL INCHES

Millimeters	0	1	2	3	4	5	6	7	8	9
	\multicolumn{10}{c}{INCH EQUIVALENTS}									
	0.00000	0.03937	0.07874	0.11811	0.15748	0.19685	0.23622	0.27559	0.31496	0.35433
10	0.39370	0.43307	0.47244	0.51181	0.55118	0.59055	0.62992	0.66929	0.70866	0.74803
20	0.78740	0.82677	0.86614	0.90551	0.94488	0.98425	1.02362	1.06299	1.10236	1.14173
30	1.18110	1.22047	1.25984	1.29921	1.33858	1.37795	1.41732	1.45669	1.49606	1.53543
40	1.57480	1.61417	1.65354	1.69291	1.73228	1.77165	1.81102	1.85039	1.88976	1.92913
50	1.96850	2.00787	2.04724	2.08661	2.12598	2.16535	2.20472	2.24409	2.28346	2.32283
60	2.36220	2.40157	2.44094	2.48031	2.51968	2.55905	2.59842	2.63779	2.67716	2.71653
70	2.75590	2.79527	2.83464	2.87401	2.91338	2.95275	2.99212	3.03149	3.07086	3.11023
80	3.14960	3.18897	3.22834	3.26771	3.30708	3.34645	3.38582	3.42519	3.46456	3.50393
90	3.54330	3.58267	3.62204	3.66141	3.70078	3.74015	3.77952	3.81889	3.85826	3.89763
100	3.93700	3.97637	4.01574	4.05511	4.09448	4.13385	4.17322	4.21259	4.25196	4.29133
110	4.33070	4.37007	4.40944	4.44881	4.48818	4.52755	4.56692	4.60629	4.64566	4.68503
120	4.72440	4.76377	4.80314	4.84251	4.88188	4.92125	4.96062	4.99999	5.03936	5.07873
130	5.11810	5.15747	5.19684	5.23621	5.27558	5.31495	5.35432	5.39369	5.43306	5.47243
140	5.51180	5.55117	5.59054	5.62991	5.66928	5.70865	5.74802	5.78739	5.82676	5.86613
150	5.90550	5.94487	5.98424	6.02361	6.06298	6.10235	6.14172	6.18109	6.22046	6.25983
160	6.29920	6.33857	6.37794	6.41731	6.45668	6.49605	6.53542	6.57479	6.61416	6.65353
170	6.69290	6.73227	6.77164	6.81101	6.85038	6.88975	6.92912	6.96849	7.00786	7.04723
180	7.08660	7.12597	7.16534	7.20471	7.24408	7.28345	7.32282	7.36219	7.40156	7.44093
190	7.48030	7.51967	7.55904	7.59841	7.63778	7.67715	7.71652	7.75589	7.79526	7.83463
200	7.87400	7.91337	7.95274	7.99211	8.03148	8.07085	8.11022	8.14959	8.18896	8.22833
210	8.26770	8.30707	8.34644	8.38581	8.42518	8.46455	8.50392	8.54329	8.58266	8.62203
220	8.66140	8.70077	8.74014	8.77951	8.81888	8.85825	8.89762	8.93699	8.97636	9.01573
230	9.05510	9.09447	9.13384	9.17321	9.21258	9.25195	9.29132	9.33069	9.37006	9.40943
240	9.44880	9.48817	9.52754	9.56691	9.60628	9.64565	9.68502	9.72439	9.76376	9.80313
250	9.84250	9.88187	9.92124	9.96061	9.99998	10.03935	10.07872	10.11809	10.15746	10.19683
260	10.23620	10.27557	10.31494	10.35431	10.39368	10.43305	10.47242	10.51179	10.55116	10.59053
270	10.62990	10.66927	10.70864	10.74801	10.78738	10.82675	10.86612	10.90549	10.94486	10.98423
280	11.02360	11.06297	11.10234	11.14171	11.18108	11.22045	11.25982	11.29919	11.33856	11.37793
290	11.41730	11.45667	11.49604	11.53541	11.57478	11.61415	11.65352	11.69289	11.73226	11.77163

AMP SPECIAL INDUSTRIES
TERMINALS AND CONNECTORS
ALL TERMINALS ACTUAL SIZE

RING TONGUE, NON-INSULATED TERMINALS

AWG SIZE 22-16

34104	34105	34108	34109	34113
4 Stud	6 Stud	8 Stud	10 Stud	¼ Stud

AWG SIZE 16-14

321684	34122	34123	34124	34126
6 Stud	8 Stud	10 Stud	¼ Stud	⅜ Stud

AWG SIZE 12-10

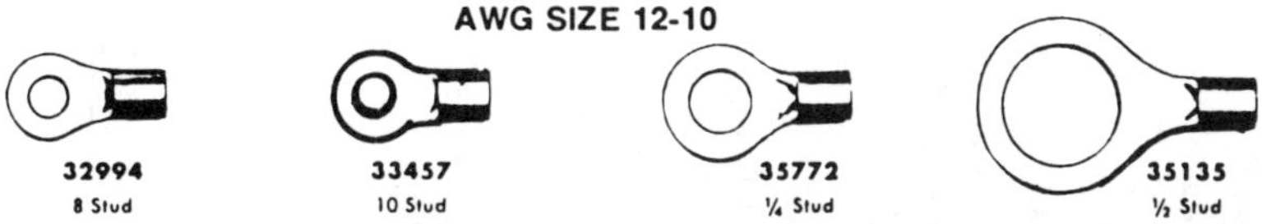

32994	33457	35772	35135
8 Stud	10 Stud	¼ Stud	½ Stud

BATTERY TERMINALS

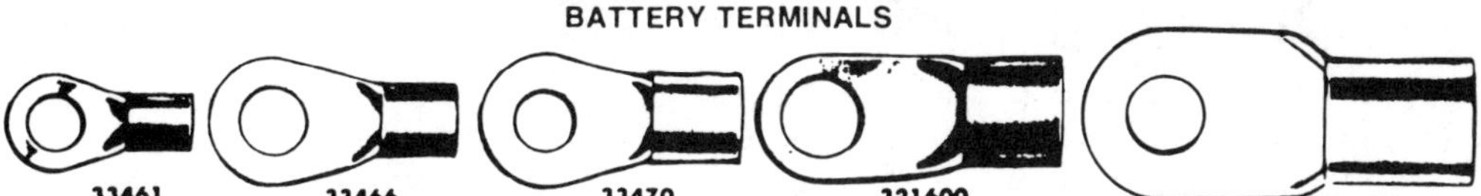

33461	33466	33470	321600	321868
¼ Stud	5/16 Stud	5/16 Stud	⅜ Stud	3/8" stud
AWG SIZE 8	AWG SIZE 6	AWG SIZE 4	AWG SIZE 2	AWG SIZE 1/0

RING TONGUE, VINYL INSULATED TERMINALS

AWG SIZE 22-16   RED

31880	31885	31890	31891	31894
4 Stud	6 Stud	8 Stud	10 Stud	¼ Stud

AWG SIZE 16-14   BLUE

32442	31902	31903	31906
6 Stud	8 Stud	10 Stud	¼ Stud

AWG SIZE 12-10   YELLOW

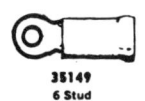

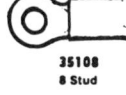

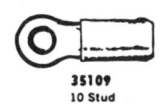

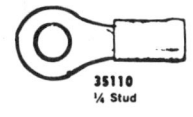

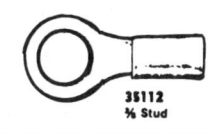

35149	35108	35109	35110	35112
6 Stud	8 Stud	10 Stud	¼ Stud	⅜ Stud

KNIFE DISCONNECTS, VINYL INSULATED

32446	32448
AWG SIZE 22-16	AWG SIZE 16-14
RED	BLUE

PLASTI-GRIP BUTT-TYPE SPLICES

34070	34071	34072
AWG SIZE 22-16	AWG SIZE 16-14	AWG SIZE 12-10
RED	BLUE	YELLOW

SELECTING FITTINGS FOR INSTRUMENT HOOKUPS

1. Make sketch of instrument layout.
2. Draw lines showing routing of tubing from instruments to the firewall or other locations.
3. Determine diameters of instrument lines. One-fourth inch diameter tubing is used for most instruments. The artificial horizon and directional gyro should have at least 3/8 inch diameter lines.
4. Select the required AN or Nylo-Seal fittings.

EXAMPLE NO. 1: Hookup of Airspeed, Altimeter, and Rate of Climb.

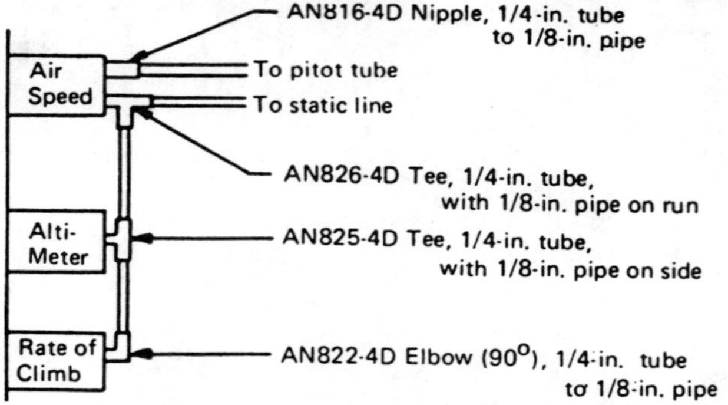

AN816-4D Nipple, 1/4-in. tube to 1/8-in. pipe

AN826-4D Tee, 1/4-in. tube, with 1/8-in. pipe on run

AN825-4D Tee, 1/4-in. tube, with 1/8-in. pipe on side

AN822-4D Elbow (90°), 1/4-in. tube to 1/8-in. pipe

The aluminum lines (1/4-in. diameter) will be flared and attached to the fittings with AN818-4D nuts and AN819-4D sleeves.

HELPFUL HINTS

It is good practice to put some form of water trap in both the pitot and static lines. It should be placed at the lowest practical point in the system between the pressure source and the instrument. The trap can be a simple T-fitting with a cap:

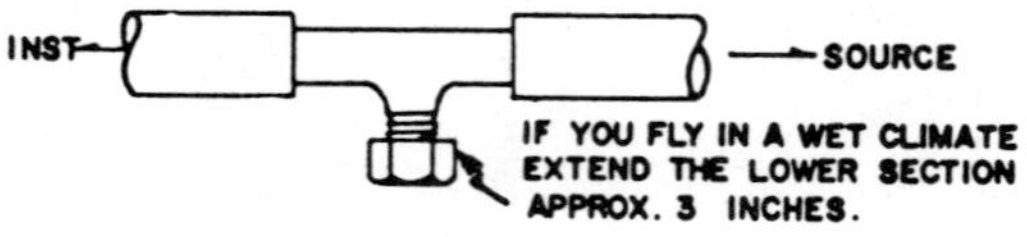

IF YOU FLY IN A WET CLIMATE EXTEND THE LOWER SECTION APPROX. 3 INCHES.

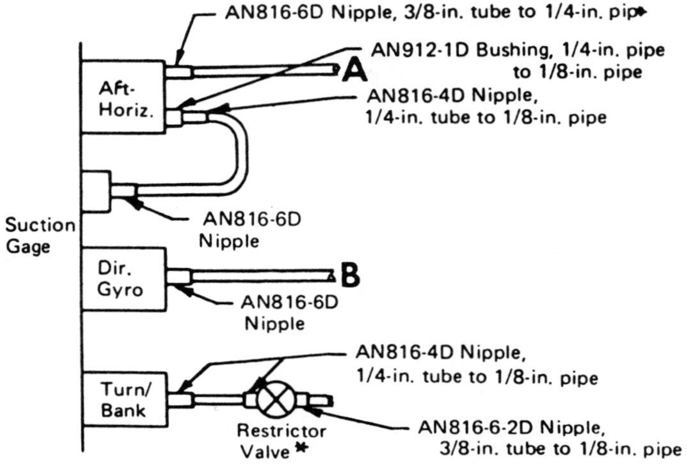

Examples of Equivalent AN-Type and Nylo-Seal Fittings

AN Fitting	Thread Size	Nylo-Seal Fitting
AN816-4D Nipple	1/4 tube x 1/8 pipe	268-N 1/4
AN816-6D "	3/8 tube x 1/4 pipe	268-N 3/8
AN818-4D Nut	1/4 tube	Nuts and Sleeves are included with all Nylo-Seal fittings.
AN818-6D "	3/8 tube	
AN819-4D Sleeve	1/4 tube	
AN819-6D "	3/8 tube	
AN822-4D Elbow	1/4 tube x 1/8 pipe	269-N 1/4 x 1/8
AN825-4D Tee	1/4 tube, 1/8 pipe on run	272-N 1/4 x 1/8
AN826-4D "	1/4 tube, 1/8 pipe on side	271-N 1/4 x 1/8
AN912-1D Bushing	1/4 pipe x 1/8 pipe	No Equivalent

* Use AN816 Nipple and Nyloseal 266-N Female Connector to transition from AN-type fittings to Nyloseal fittings.

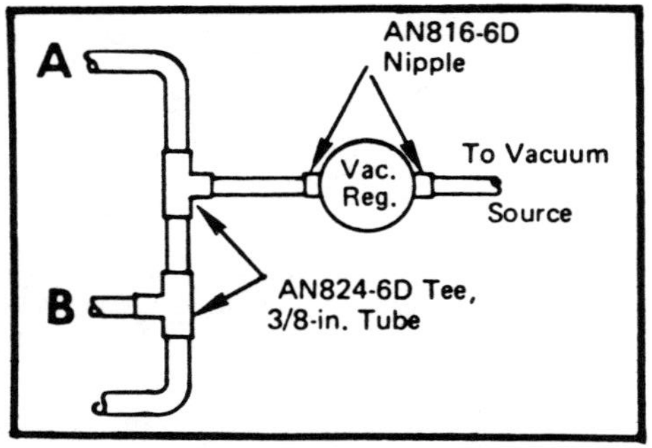

EXAMPLE NO. 2: Hookup of Gyro Instruments.

Each connection between the aluminum lines and the fittings requires either an AN818-4D nut and AN819-4D sleeve (1/4-in. line) or an AN818-6D nut and an AN819-6D sleeve for 3/8-in. line.
*Turn/Bank vacuum set at 2 In.Hg.

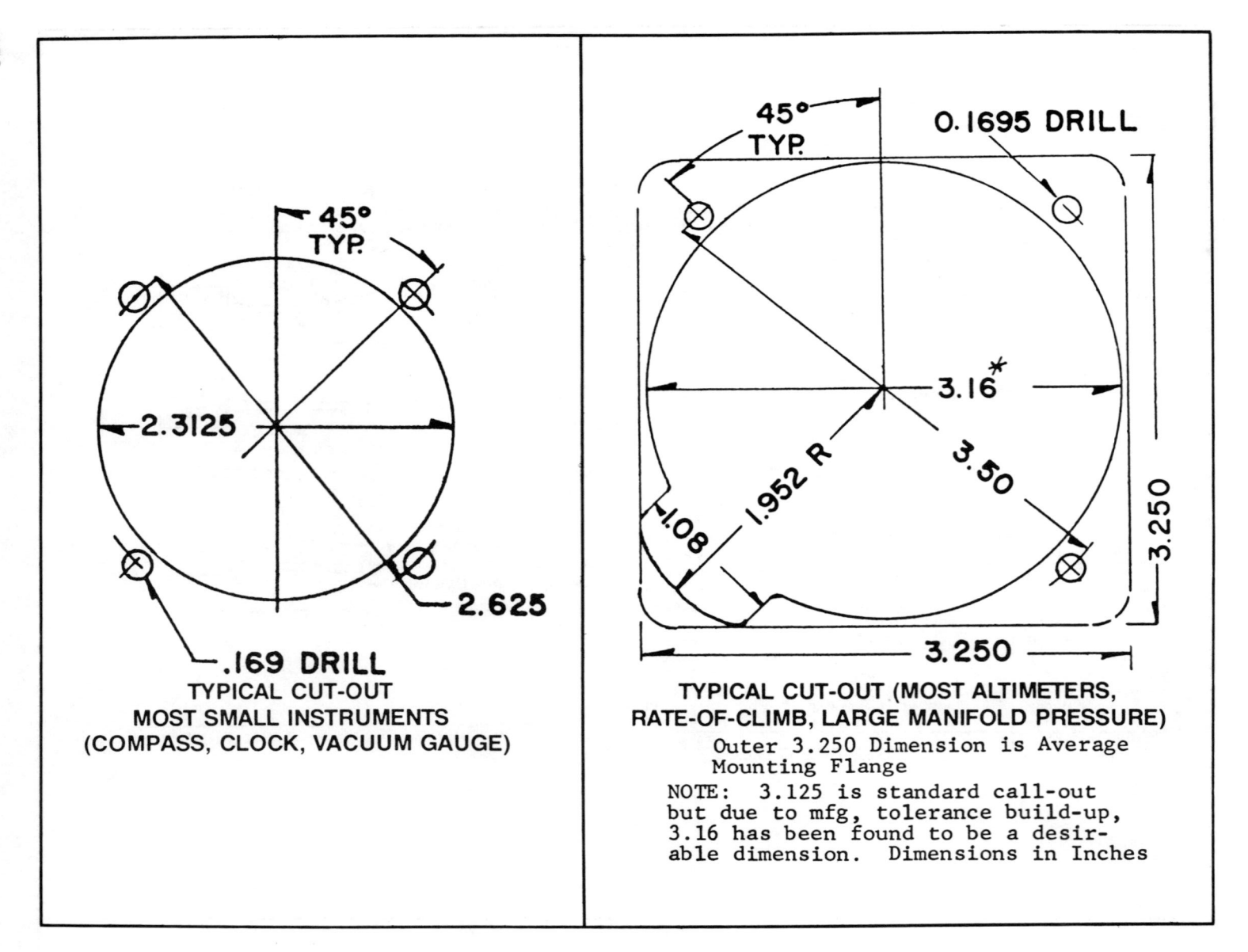

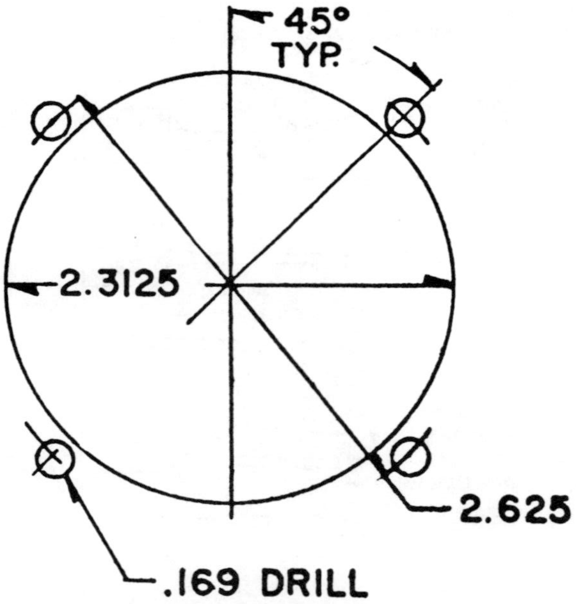
45°
TYP.
2.3125
2.625
.169 DRILL

TYPICAL CUT-OUT
MOST SMALL INSTRUMENTS
(COMPASS, CLOCK, VACUUM GAUGE)

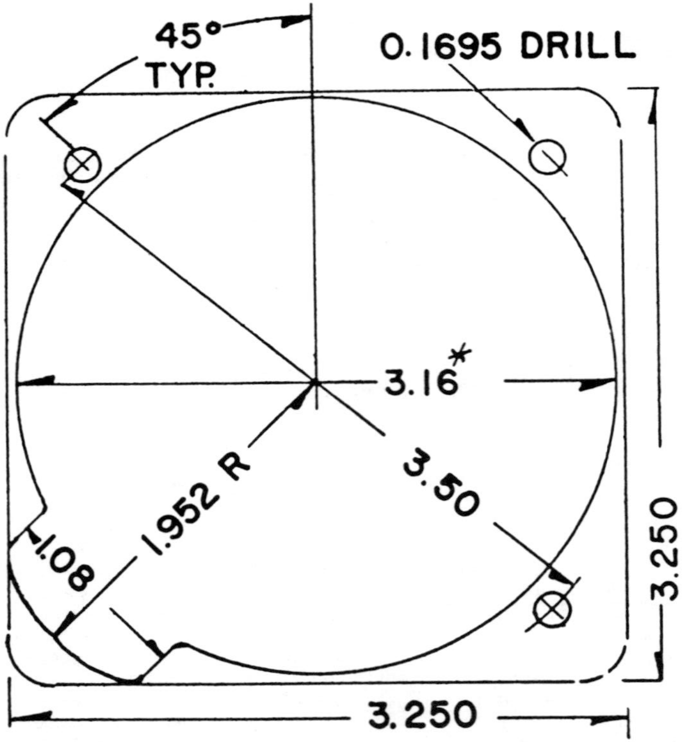
45°
TYP.
0.1695 DRILL
3.16*
3.50
1.952 R
1.08
3.250
3.250

TYPICAL CUT-OUT (MOST ALTIMETERS, RATE-OF-CLIMB, LARGE MANIFOLD PRESSURE)

Outer 3.250 Dimension is Average Mounting Flange

NOTE: 3.125 is standard call-out but due to mfg, tolerance build-up, 3.16 has been found to be a desirable dimension. Dimensions in Inches

FULL-SCALE INSTRUMENT CUT-OUTS

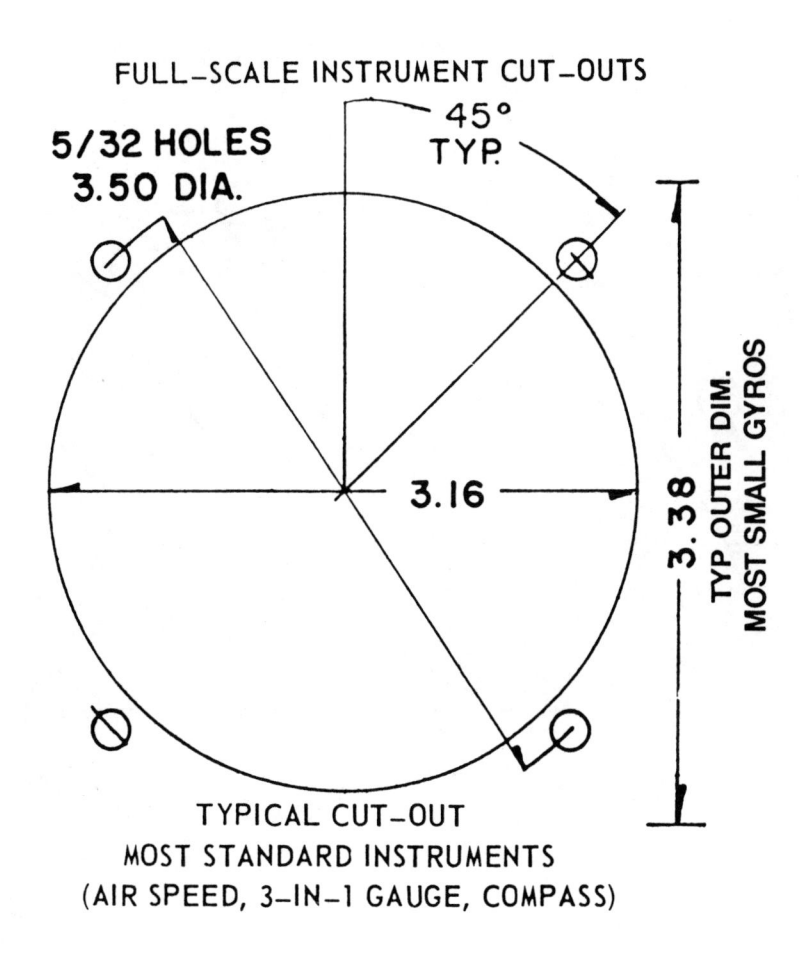

5/32 HOLES
3.50 DIA.

45° TYP.

3.16

3.38 TYP OUTER DIM. MOST SMALL GYROS

TYPICAL CUT-OUT
MOST STANDARD INSTRUMENTS
(AIR SPEED, 3-IN-1 GAUGE, COMPASS)

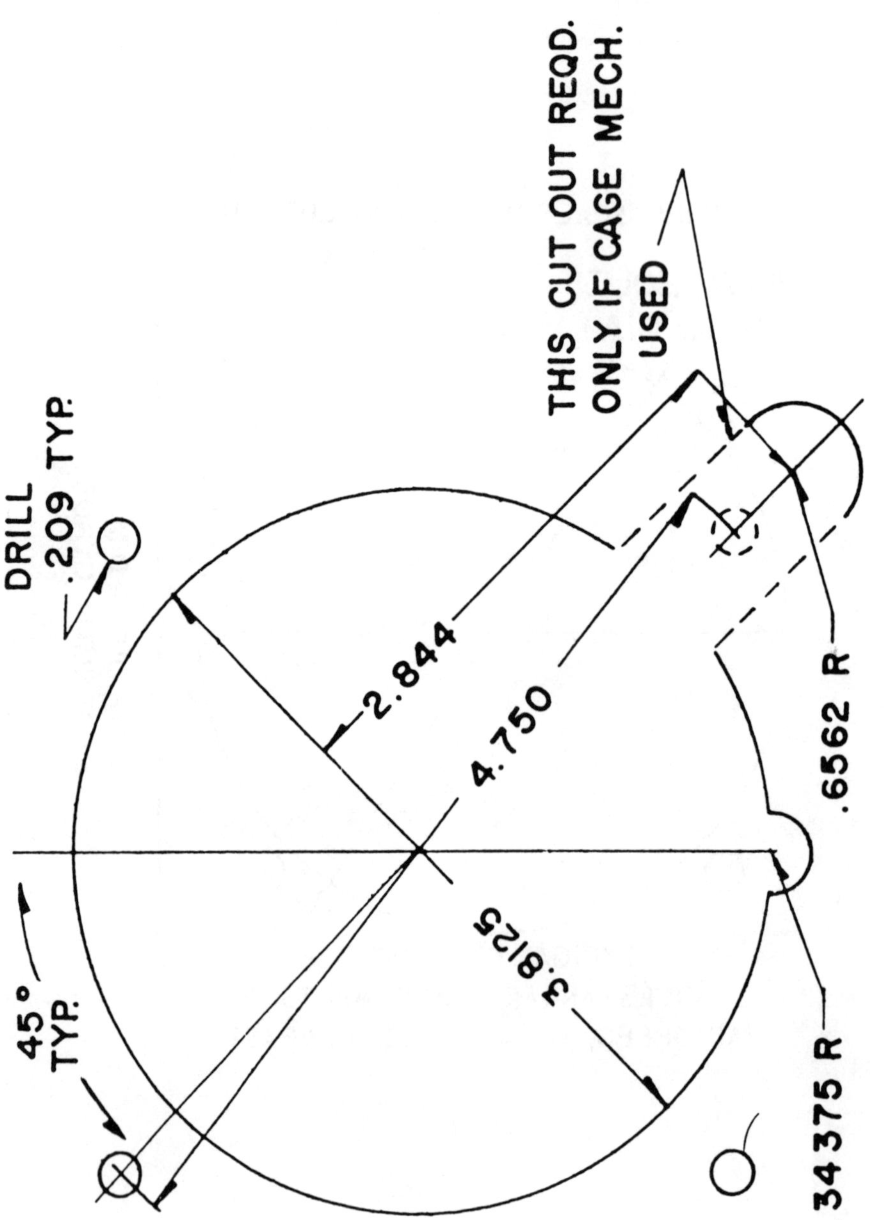
DRILL .209 TYP.
45° TYP.
2.844
4.750
3.8125
.6562 R
.34375 R
THIS CUT OUT REQD. ONLY IF CAGE MECH. USED

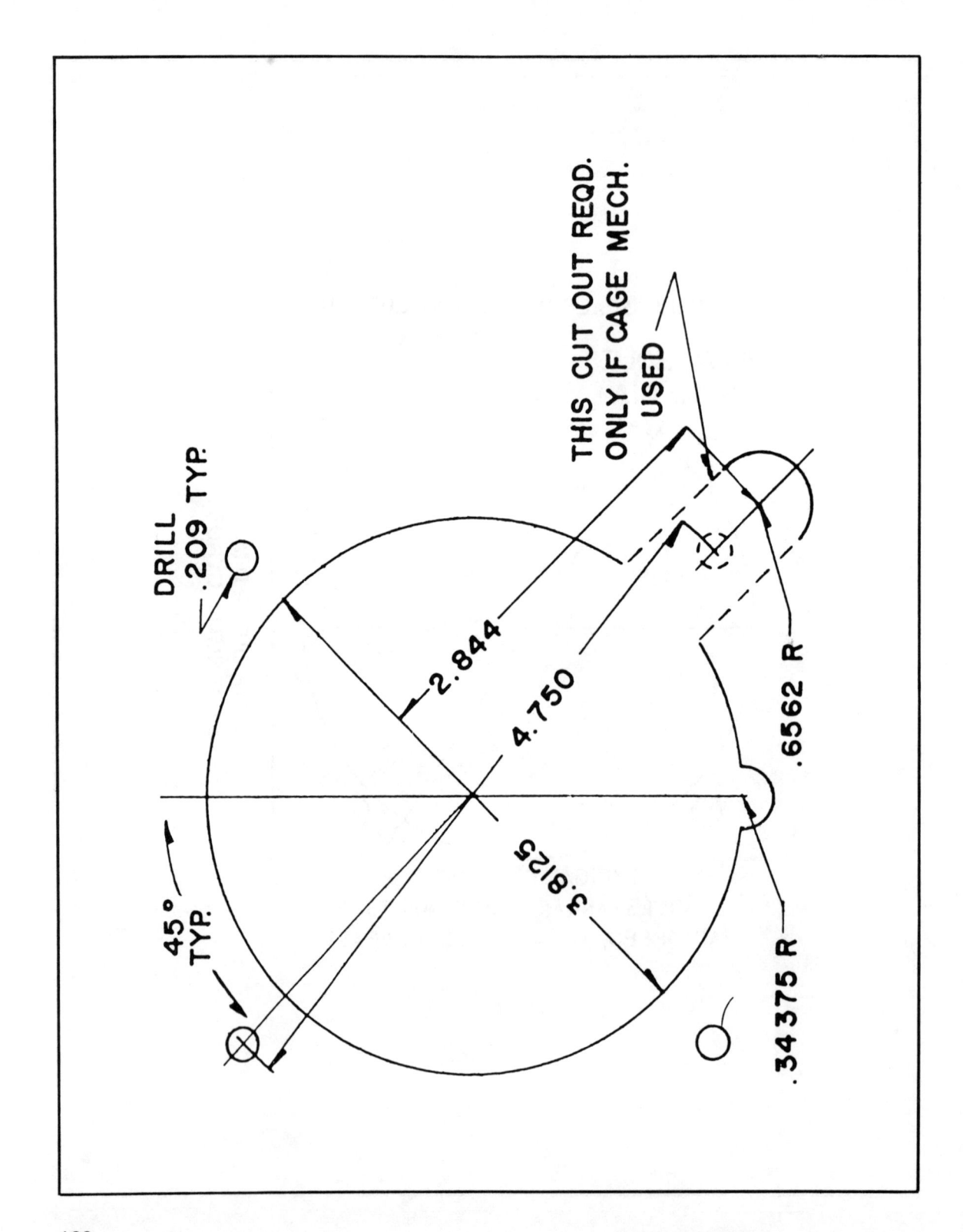

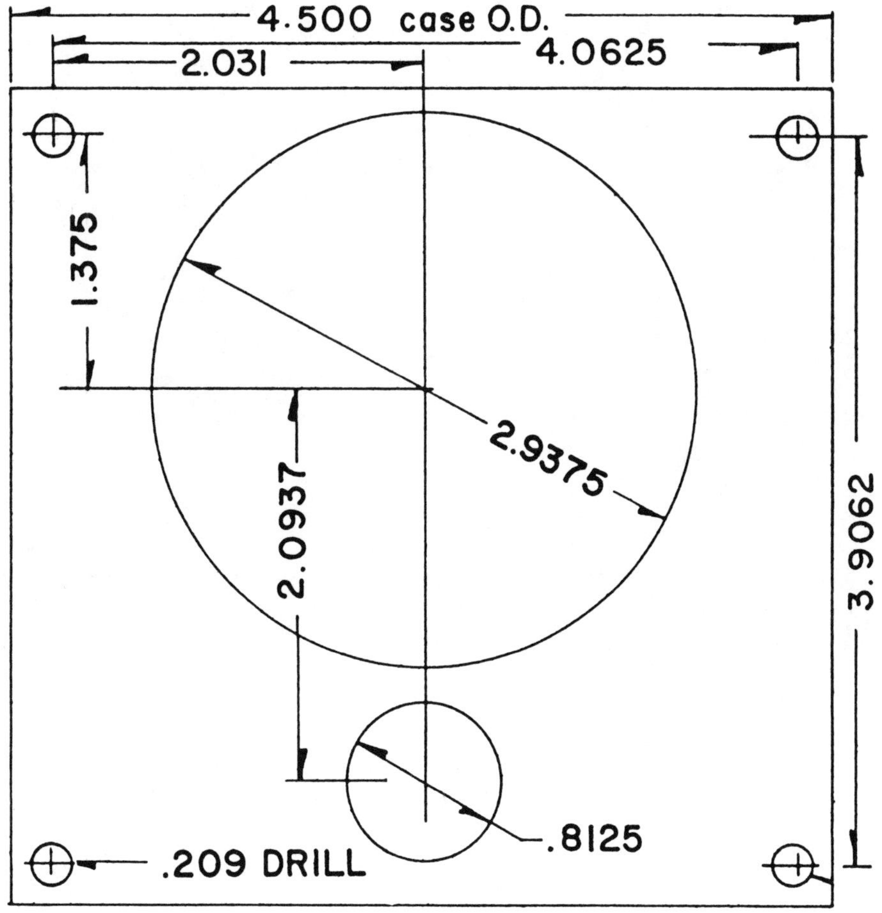
4.500 case O.D.
2.031
4.0625
1.375
2.9375
2.0937
3.9062
.8125
.209 DRILL

Index

Index

A
Aero Mirage TC-2, 73-74
allergies, 133
all-metal construction, 1
AMI engines, 31-32
antennas, 65-72
AVCO Lycoming engines, 26-27
avionics, 52-72

B
Beechcraft A36, 1
bidirectional cloth, 4

C
carbon fibers, 7
ceramic cloth, 5, 7
Cessna 152, 1
chrome plating, 22
codes, engine, 23
composite, definition of, 1
Cozy, 74-76
crowfoot weave fiberglass, 7
dermatitis, 133
Dragonfly, 76-79
Duncan rotary engine, 43-47

E
E-Glass, 5
engine codes, 23
engine storage, 28-30
engines, certified, 21-30
engines, non-certified, 31
Experimental Aircraft Association (EAA), 121-130

F
FAA regulations, 100-120
fiberglass tape, 12
fiberglass, crowfoot weave, 7
fiberglass, plain weave, 7
fiberglass, unidirectional, 7-8
foam-over-structure, 13-14

G
Glasair, 79-81
glass mat, 8
GPASC engines, 43
graphite fiber, 5
graphite, unidirectional, 6-7
graphite, unidirectional woven, 6

H
H.A.P.I. engines, 35-43
health, 133

I
INAV Ltd, 36-37
instrumentation, 72

K
Kevlar, 5-6
KS-400, 6

L
Lancer, 81-83
Limbach Flugmotoren engines, 32-35

M
Magnaflux/Magnaglow, 22
Micro-Putty, 12
moldless foam structure, 14-15
Monnett engines, 36-37

N
nitriding, 22

P
Peel Ply, 12
Piper J3, 1
Piper Tri-Pacer, 1
plain weave fiberglass, 7
Polliwagen, 83-84
practice kits, 131
premolded structure, 15
propellers, 49-51
purchasing, 133-134

Q
quality control, 131
Quickie/Q2, 84-87

R
radio kits, 65
Rand KR-1/KR-2, 87-90
rotary engines, 43-47
Rutan fiberglass cloth, 4
Rutan VariEze/Long-EZ/Defiant/Solitaire, 90-94

S
safety, flying, 134
Sea Hawk, 94-95
Sensenich Corp., 49-51
S-Glass, 5, 7
Silhouette, 95-97
Solo, 97
Star-Lite, 97-99
Sterling filler, 12

T
TBO, 22
Teledyne Continental engines, 23-26
3M #77 spray adhesive, 12
top overhaul, 21-22
tube-and-fabric construction, 1

U
unidirectional cloth, 4
unidirectional fiberglass, 7-8
unidirectional woven graphite, 6

V
value, engine, 22-23
VW conversion engines, 35-43

W
W.A.R. replicas, 99

OTHER POPULAR TAB BOOKS OF INTEREST

Unconventional Aircraft (No. 2384—$17.50 paper)
Celluloid Wings (No. 2374—$25.50 paper)
Guide to Homebuilts—9th Edition (No. 2364—$11.50 paper; $17.95 hard)
Flying Hawaii—A Pilot's Guide to the Islands (No. 2361—$10.25 paper)
Flying in Congested Airspace (No. 2358—$10.25 paper)
How to Master Precision Flight (No. 2354—$9.95 paper)
Fun Flying!—A Total Guide to Ultralights (No. 2350—$10.25 paper)
Engines for Homebuilt Aircraft & Ultralights (No. 2347—$8.25 paper)
Student Pilot's Solo Practice Guide (No. 2339—$12.95 paper)
The Complete Guide to Homebuilt Rotorcraft (No. 2335—$7.95 paper)
Pilot's Guide to Weather Forecasting (No. 2331—$9.25 paper)
Flying The Helicopter (No. 2326—$10.95 paper)
The Beginner's Guide to Flight Instruction (No. 2324—$13.50 paper)
How To Become A Flight Engineer (No. 2318—$6.95 paper)
Aircraft Dope and Fabric—2nd Edition (No. 2313—$9.25 paper)
How To Become An Airline Pilot (No. 2308—$8.95 paper)
Building & Flying the Mitchell Wing (No. 2302—$7.95 paper)
Instrument Flying (No. 2293—$8.95 paper)
The ABC's of Safe Flying (No. 2290—$8.25 paper; $12.95 hard)
Cross-Country Flying (No. 2284—$8.95 paper; $9.95 hard)
The Complete Guide to Single-Engine Cessnas—3rd Edition (No. 2268—$7.95 paper; $10.95 hard)

Aircraft Construction, Repair and Inspection (No. 2377—$13.50 paper)
Your Pilot's License—3rd Edition (No. 2367—$9.95 paper)
Welcome to Flying: A Primer for Pilots (No. 2362—$13.50 paper)
The Complete Guide to Rutan Aircraft—2nd Edition (No. 2360—$13.50 paper)
Build Your Own Low-Cost Hangar (No. 2357—$9.25 paper)
Sabre Jets Over Korea: A Firsthand Account (No. 2352—$15.50 paper)
All About Stalls and Spins (No. 2349—$9.95 paper)
The Pilot's Health (No. 2346—$15.50 paper)
Your Mexican Flight Plan (No. 2337—$12.95 paper)
The Complete Book of Cockpits (No. 2332—$39.95 hard)
American Air Power: The First 75 Years (No. 2327—$21.95 hard)
The Private Pilot's Handy Reference Manual (No. 2325—$11.50 paper; $14.95 hard)
The Joy of Flying (No. 2321—$10.25 paper)
The Complete Guide to Aeroncas, Citabrias and Decathlons (No. 2317—$15.50 paper)
Lift, Thrust & Drag—a primer of modern flying (No. 2309—$8.95 paper)
Aerial Banner Towing (No. 2303—$7.95 paper)
31 Practical Ultralight Aircraft You Can Build (No. 2294—$8.25 paper)
Your Alaskan Flight Plan (No. 2292—$8.95 paper)
Pilot's Weather Guide—2nd Edition (No. 2288—$7.95 paper)
Flying VFR in Marginal Weather (No. 2282—$9.25 paper)
How to Fly Helicopters (No. 2264—$10.25 paper)
How to Take Great Photos from Airplanes (No. 2251—$5.95 paper; $8.95 hard)

TAB TAB BOOKS Inc.

Blue Ridge Summit, Pa. 17214

Send for FREE TAB Catalog describing over 750 current titles in print.